SAMS Teach Yourself
MICROSOFT® PUBLISHER 2000

Joe Habraken

in 10 Minutes

SAMS

A Division of Macmillan Computer Publishing
201 West 103rd St., Indianapolis, Indiana, 46290 USA

To my cousin, Mary, and her husband, Tony. You both make the world a better place.

Copyright © 1999 by Sams Publishing

All rights reserved. No part of this book shall be reproduced, stored in a retrieval system, or transmitted by any means, electronic, mechanical, photocopying, recording, or otherwise, without written permission from the publisher. No patent liability is assumed with respect to the use of the information contained herein. Although every precaution has been taken in the preparation of this book, the publisher and author assume no responsibility for errors or omissions. Neither is any liability assumed for damages resulting from the use of the information contained herein.

International Standard Book Number: 0-672-31569-6

Library of Congress Catalog Card Number: 98-89862

Printed in the United States of America

First Printing: March 1999

01 00 99   4 3 2 1

TRADEMARKS

All terms mentioned in this book that are known to be trademarks or service marks have been appropriately capitalized. Sams cannot attest to the accuracy of this information. Use of a term in this book should not be regarded as affecting the validity of any trademark or service mark.

WARNING AND DISCLAIMER

Every effort has been made to make this book as complete and as accurate as possible, but no warranty or fitness is implied. The information provided is on an "as is" basis. The authors and the publisher shall have neither liability nor responsibility to any person or entity with respect to any loss or damages arising from the information contained in this book.

EXECUTIVE EDITOR
Jim Minatel

ACQUISITIONS EDITOR
Jenny Watson

DEVELOPMENT EDITOR
Jill Hayden

MANAGING EDITOR
Brice Gosnell

PROJECT EDITOR
Gretchen Uphoff

COPY EDITOR
Kelly Talbot

INDEXER
John Jefferson

PROOFREADER
Benjamin Berg

TECHNICAL EDITOR
Coletta Witherspoon

INTERIOR DESIGN
Gary Adair

COVER DESIGN
Aren Howell

LAYOUT TECHNICIANS
Brian Borders
Heather Moseman
Mark Walchle

Contents

Introduction   **1**

The What and Why of Microsoft Publisher 1
Why Sams Teach Yourself Yourself Microsoft Publisher 2000 in 10 Minutes? 2
Installing Publisher 2
Conventions Used in This Book 4

1 What's New in Publisher 2000   **5**

Getting to Know Publisher 5
New Features in Publisher 2000 6

2 Getting Started with Publisher   **11**

Starting Publisher 11
Deciding How to Create a New Publication 15
Using Menus and Toolbars 17
Understanding Dialog Boxes 21
Exiting Publisher...22

3 Creating a New Publication   **24**

Planning Your Publication 24
Using the Publication Wizards 25
Saving Your Publication 32

4 Using Design Sets and Templates   **34**

Understanding the Publication Design Sets 34
Selecting the Design Set 34
Completing the Publication 37
Adding Your Own Design and Color Schemes 38

5 Viewing Your Publications   **42**

Changing the Publication Display 42
Using the Zoom Feature 45
Scrolling in the Publication 46
Working with Rulers and Guide Lines 47

6 Working with Existing Publications 52

Opening an Existing Publication 52
Completing a Wizard-Based Publication 54
Adding Pages to a Publication 57
Saving a Revised Document Under a New Name 59
Closing a Publication 60

7 Getting Help in Publisher 61

Using the Office Assistant 61
Getting Help Without the Assistant 63
What's This? 68
Getting Help Online 68

8 Working with Publication Frames 69

Inserting a Frame 69
Sizing a Frame 71
Moving a Frame 74
Copying a Frame 76
Grouping Frames 76
Arranging Frames in Layers 78

9 Enhancing Frames with Borders and Colors 80

Adding Borders to Frames 80
Changing Border Attributes 81
Using Fill Colors 83
Using Fill Effects 84
Applying Shading 85

10 Changing How Text Looks 87

Adding Text to Your Publications 87
Working with Fonts 89
Changing Font Attributes 91
Changing Font Colors 92
Aligning Text in a Frame 93
Adding Text Mastheads 94
Connecting Text Frames 96

11 Working with Graphics — 99
Inserting a Picture 99
Using Clip Art 101
Scaling Pictures 104
Cropping Pictures 105
Changing Picture Colors 106

12 Adding Special Objects to Your Publications — 108
Using the Design Gallery 108
Editing Design Gallery Objects 109
Inserting Objects from Other Applications 111
Inserting Video and Audio 113
Acquiring Images from Scanners and Other Sources 116

13 Drawing Objects in Publisher — 118
Using the Drawing Tools 118
Formatting Drawing Objects 122
Rotating an Object 123
Drawing with Microsoft Draw 124

14 Working with Line Spacing, Indents, and Lists — 127
Setting Line Spacing in a Text Frame 127
Indenting Text 128
Setting Tabs 129
Working with Numbered Lists 131
Adding Bullets to Your Text Lists 132

15 Working with Publication Tables — 134
Inserting a Table 134
Sizing and Moving Tables 135
Sizing Table Columns and Rows 137
Adding Columns and Rows to the Table 138
Using Special Cell Formats 139
Filling Your Table with Information 141
Formatting the Table Automatically 142
Formatting the Table Manually 143

16 Formatting Publication Pages — 145
Changing Page Margins 145
Adding Page Borders 146
Working in the Publication Background 148

17 Fine-Tuning Publisher Publications — 153
Using the Spell Checker 153
Controlling Hyphenation in Text Frames 155
Using the Design Checker 156
Setting Up AutoCorrect 158

18 Printing and Outputting Publisher Publications — 161
Previewing the Publication 161
Printing the Publication 162
Working with Print Options 163
Troubleshooting Printing Problems 166
Working with an Outside Print Service 167
Using Pack and Go 170

19 Mass Mailing Publications — 172
Understanding the Mail Merge Feature 172
Building a Mailing List...173
Starting the Merge and Inserting Merge Codes 175
Completing the Merge 177

20 Creating Publications on Special Paper — 180
Creating Trifold Brochures...180
Creating Business Cards...182
Creating Envelopes 183
Creating Mailing Labels 185

21 Creating a Publisher Web Site — 187
What Is the World Wide Web? 187
Creating a Web Site Using the Web Site Wizard 188
Converting an Existing Publication to a Web Site 191
Adding and Removing Hyperlinks 193
Viewing Your Web Site 195
Publishing Your Web Site 196

Acknowledgments

Creating books like this takes a real team effort. I would like to thank Jenny Watson, our acquisitions editor, who worked very hard to assemble the team that made this book a reality. I would also like to thank Jill Hayden, who served as the development editor for this book and who came up with many great ideas for improving the content of the book. Also a tip of the hat and a thanks to Coletta Witherspoon, who as the technical editor for the project did a fantastic job making sure that everything was correct and suggested a number of additions that made the book even more technically sound. Finally, a great big thanks to our project editor, Gretchen Uphoff, who ran the last leg of the race and made sure the book made it to press on time—what a great team of professionals.

TELL US WHAT YOU THINK!

As the reader of this book, *you* are our most important critic and commentator. We value your opinion and want to know what we're doing right, what we could do better, what areas you'd like to see us publish in, and any other words of wisdom you're willing to pass our way.

As an associate publisher at Sams Publishing, I welcome your comments. You can fax, email, or write me directly to let me know what you did or didn't like about this book—as well as what we can do to make our books stronger.

Please note that I cannot help you with technical problems related to the topic of this book, and that due to the high volume of mail I receive, I might not be able to reply to every message.

When you write, please be sure to include this book's title and author as well as your name and phone or fax number. I will carefully review your comments and share them with the author and editors who worked on the book.

Fax:	317.581.4770	
Email:	office_que@mcp.com	
Mail:	Associate Publisher	
	Sams Publishing	
	201 West 103rd Street	
	Indianapolis, IN 46290 USA	

Introduction

Microsoft Publisher is a desktop publishing tool that is easy-to-use but can help you to build very sophisticated publications. You can create print publications, such as newsletters, greeting cards, and brochures, and you can create online publications, such as Web pages for your Web site. You can choose to print your own publications or to have Publisher help you prepare your publication for printing by a professional print service.

Publisher also provides you with a number of wizards that walk you through the publication process, making it easy for you to select design elements, color schemes, and publication layouts that make your publications look professional. Publisher is tightly integrated with the other applications in Microsoft Office 2000, making it easy for you to include objects in your publications that were created in Microsoft Word, Excel, or PowerPoint.

The What and Why of Microsoft Publisher

Publisher can help you create a wide variety of publication types. It supplies you with a great deal of help and a number of special tools for creating your publications:

- Specific wizards (such as the Business Card Wizard) help you decide on the overall publication design for a particular publication type.
- The Clip Gallery provides you with a large number of clip art images for use in your publications.
- The Design Gallery supplies you with special design elements such as mastheads, reply forms, and logos that you can add to your publication pages.
- A set of drawing tools makes it easy for you to create your own artwork and design elements for your publications.

Additionally, Publisher looks and operates like all the other Office 2000 applications. They embrace a common interface and help system that enable you to do the following:

- Use the Office Assistant to get quick help
- Take advantage of toolbar buttons to quickly fire off Publisher commands
- Easily import items from other Office applications

Microsoft Publisher is easy to learn and offers many advantages and benefits in return. This book can help you understand the possibilities awaiting you with Microsoft Publisher.

This book concentrates on using Publisher on a Windows 98 (or 95) workstation on which Microsoft Office is also installed. Note, however, that you can also install Microsoft Publisher on a computer running Windows NT 4.0.

Why *Sams Teach Yourself Yourself Microsoft Publisher 2000 in 10 Minutes*?

Sams Teach Yourself Microsoft Publisher 2000 in 10 Minutes can save you precious time while you get to know the program. Each lesson is designed to be completed in 10 minutes or less, so you'll be up to snuff in basic Publisher skills quickly.

Although you can jump around among lessons, starting at the beginning is a good plan. The bare-bones basics are covered first, and more advanced topics are covered later. If you need help installing Publisher, see the next section for instructions.

Installing Publisher

You can install Microsoft Publisher on a workstation running Windows 95, Windows 98, or Windows NT 4.0. (Publisher does *not* run on a computer running Windows for Workgroups, Windows 3.x, or Windows NT 3.5.) In addition, you can install Publisher in conjunction with Microsoft Office 2000, or you can install just the Publisher program.

To install Publisher, follow these steps:

1. Start your computer. Then insert the Microsoft Office CD 2 (or the Microsoft Publisher CD, depending on whether you purchased Publisher as a stand-alone product or as part of the Office suite) in the CD-ROM drive. The Office installation program should start automatically.

2. If the Office Installation program does not appear on your computer's desktop, select Start, Run. In the Run dialog box, type the letter of the CD-ROM drive, followed by **setup** (for example, **e:\setup**). If necessary, use the Browse button to locate and select the CD-ROM drive and the setup.exe program.

3. Whether the installation program starts automatically or you start it, you then are prompted to provide your name and organization. Do this and click Next to continue.

4. The next step in the installation process asks you for the CD-ROM key. You can usually find it on a sticker somewhere on the jewel case that the CD-ROM came in. Type in the number and click Next to continue.

5. Follow the remaining onscreen instructions to choose a location for your Publisher installation and to select the software component you want to install.

The new Microsoft Installation Interface lists an icon for each of the Office products available on your CD, such as Publisher. A plus symbol next to a particular software application signifies that you can open and view all the components for that application. You have the option of clicking a particular option and then choosing from a menu that enables you to select how you want the component installed: Run From My Computer (meaning it is installed on your PC), Run From CD (the component is run from CD, so make sure you keep it in the CD-ROM drive), and Installed on First Use (the component is not installed from the CD until you use the component for the first time).

After you complete the installation from the CD, you are ready to run your Office applications, such as Publisher 2000.

When you start Publisher for the first time, the Office Assistant appears. To begin using Publisher, click Start Using Publisher in the Office Assistant's balloon.

Microsoft offers software upgrades via their Web site. You can download updates and fixes for Publisher and the other Microsoft Office applications. Go to http://www.microsoft.com/Office. Use the search feature on this page to locate additional information and updates related to Publisher.

Conventions Used in This Book

To help you move through the lessons easily, these conventions are used:

Onscreen text Onscreen text and keys you are to press appear in bold type.

Items you select Commands, options, and icons you are to select appear in colored type.

In addition to those conventions, the *Sams Teach Yourself Microsoft Publisher 2000 in 10 Minutes* uses the following icons to identify helpful information:

Tips Read these tips for ideas that cut corners and confusion.

Caution This icon identifies areas where new users often run into trouble; this icon offers practical solutions to those problems.

Plain English New or unfamiliar terms are defined in (you got it) "plain English."

Lesson 1
What's New in Publisher 2000

In this lesson, you are introduced to Publisher's desktop publishing features, and you learn what's new in Publisher 2000.

Getting to Know Publisher

Publisher 2000 is the latest version of Microsoft's popular and powerful desktop publishing tool. Publisher enables you to create a large range of publications, everything from brochures to banners to business cards. Publisher's various features and tools also make it easy to produce these publications.

Publisher actually provides three different routes for the creation of the different publication types:

- **Wizards**   Wizards are provided for nearly every publication type you can think of. The wizards walk you through the steps of creating the publication and also provide you a great deal of flexibility in customizing your publication (for more about creating publications using a wizard, see Lesson 3, "Creating a New Publication").

- **Design Sets**   You can create a publication that uses the same designs and layout attributes as other publications in the same design family. Wizards are available to help you through the process of using design sets, and you can create a complete group of design-integrated publications for your small business. For more about design sets, see Lesson 4, "Using Design Sets and Templates."

- **Publications from Scratch**   For maximum flexibility, you can create your publications from a blank publication. Publisher provides help when you create this type of publication as well, and you can use the Quick Publication Wizard to help you with color and design decisions. For more about working with blank publications, see Lesson 4.

Whichever route you take, Publisher makes it easy for you to create and fine-tune any publication.

NEW FEATURES IN PUBLISHER 2000

Publisher 2000 contains a number of improvements that make it even easier for you to create great-looking publications. A number of these features are a result of a tighter integration of Publisher into the Microsoft Office family. This newest version of Publisher looks and acts a lot like other Office 2000 applications, such as Microsoft Word.

Other new features such as the Quick Publications Wizard (mentioned in the previous section) and the new personalized menu system are designed to enable you to have control over the environment that you create your publications in.

THE PERSONALIZED MENU AND TOOLBAR SYSTEM

Publisher and the other software members of Office 2000 have adopted a new menu system that provides you the capability to quickly access the commands that you use most often. When you first choose a particular menu, you find a short list of menu commands. As you use commands, Publisher adds them to the Menu list.

If you want to view all the commands available for the menu system, follow these steps:

1. Click the Tools menu, and then click Customize.

2. Click the Options tab on the Customize dialog box.

3. To show all the commands on the menus, clear the Menus show recently used commands first check box.

This personalized strategy is also embraced by the Formatting Toolbar. As you use commands, they are added to the toolbar. This provides you with customized menus and toolbars that are personalized for you. For more about using menus and toolbars, see Lesson 2, "Getting Started with Publisher."

AUTOMATIC NUMBERING AND BULLETS

This version of Publisher makes it much easier to add numbers and bullets to text lists. The formatting toolbar offers both a Numbering button and a Bullets button that can be used to quickly add either numbers or bullets to selected text respectively (see Figure 1.1).

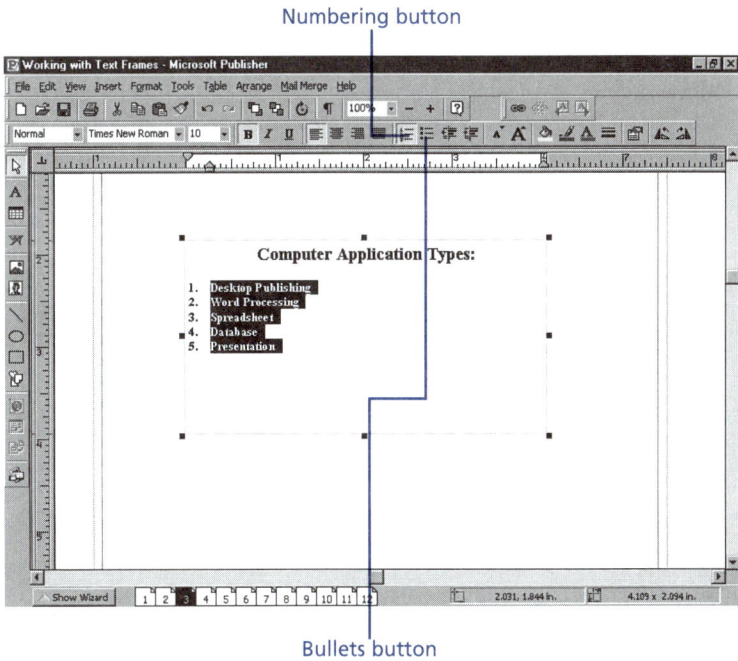

FIGURE 1.1 A Numbering button and a Bullets button make it easy for you to create numbered or bulleted lists.

For more about using automatic numbering and creating bulleted lists, see Lesson 14, "Working with Line Spacing, Tabs, and Numbered Lists."

THE PUBLISHER HELP SYSTEM

The Publisher Help system now has a very different look when compared to previous versions. Even the Office Assistant has cast aside its Assistant's box and now resides directly on the Publisher application window.

In addition, the Office Assistant is more intuitive then ever. When the Assistant displays a light bulb over its head, click the Assistant to receive help on the feature or action you are currently working with.

If you never really warmed up to the Office Assistant, you can now easily turn off this help feature. Just clear the Use Office Assistant check box on the Assistant's Options dialog box. The help system can then be accessed directly from the Help menu on the Outlook menu bar.

When you choose not to use the Office Assistant, you access a new and improved Help system. The Publisher Help system provides you with an environment that is like what you find in a Web browser. Help topics are represented as hyperlinks, and Back and Forward buttons make it easy for you to move backward and forward through the help screens that you access (see Figure 1.2).

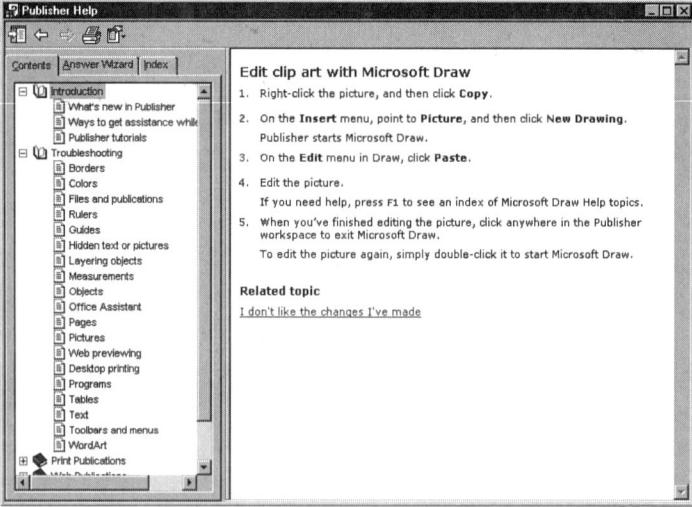

Figure 1.2 The new Help system provides you with an easy-to-use environment.

The Help feature now contains an Answer Wizard that can answer questions you put to the Help system. It works much the same as the Office Assistant. The Answer Wizard tab, along with the Contents tab and Index tab, provides you with a set of powerful choices for getting help in Publisher. For more about getting help in Publisher see Lesson 7, "Getting Help in Publisher."

OTHER NEW FEATURES

Publisher 2000 contains a number of other new features. It also improves on features that were available in previous editions of the software. These features include the following:

- **Flip Horizontal/Vertical**   You can now flip objects horizontally or vertically (see Lesson 13, "Drawing Objects in Publisher," for more information).

- **Web Site Wizard**   A new wizard that helps you create your own Web site (see Lesson 21, "Creating a Publisher Web Site," for more information).

- **Save as Web Page**   You can save an existing publication as a Web page in the HTML format (see Lesson 21 for more information).

- **Pack and Go Wizard**   A tool that helps you compress your publication on a disk so that you can take it to another computer or a commercial printer (see Lesson 18, "Printing and Outputting Publisher Publications," for more information).

- **Commercial printing support**   A set of tools that help you get your publication ready for a commercial printer (see Lesson 18 for more information).

Features that have been improved in this version of Publisher include the Design Checker, the Publisher Catalog, and the Design Gallery. Publisher 2000 has had quite a makeover from earlier versions. It is easier to use yet provides you with the capability to create even more sophisticated publications.

In this lesson, you learned how Publisher 2000 helps you create your publications. You also learned about new features and improvements in this version of Publisher. In the next lesson, you will take a first look at the Publisher application window and learn how to start and exit the software. You will also learn to decide the route you want to take to create a new publication, and you will work with the menu and toolbar systems.

Lesson 2
Getting Started with Publisher

In this lesson, you learn how to start Publisher and how to use such common tools as menus, toolbars, and dialog boxes. You also learn the process of planning a new publication and preview the three different options for creating a publication.

Starting Publisher

Publisher makes it easy for you to create a variety of different publication types. These publications can range from business cards to tri-fold brochures to Web pages. However, before you can take advantage of Publisher's sophisticated, easy-to-use tools for creating great-looking publications, you need to open the Publisher application window.

To start the Publisher program, follow these steps:

1. From the Windows Desktop, click Start, and then select Programs. The Programs menu appears (see Figure 2.1).

2. Select Microsoft Publisher. The Publisher program window appears on the Desktop.

The Publisher window looks and feels like other Office applications such as Microsoft Word or Excel. However, whereas these applications usually present you with a blank workspace such as a blank document or blank worksheet to begin working in, Publisher opens the Publisher Catalog, which provides you with three different ways of creating new publications: Publications by Wizard, Publications by Design, and Blank Publications. These three avenues to creating your publications are discussed in the next section of this lesson.

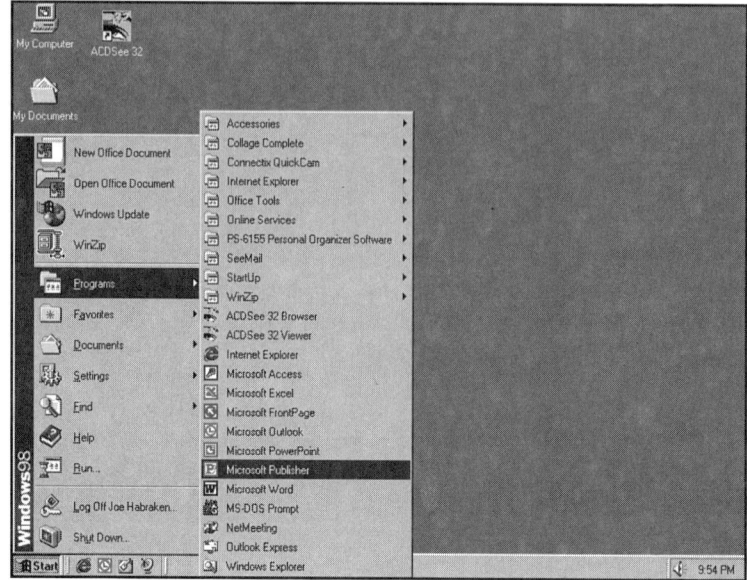

FIGURE 2.1 Open the Programs menu and select Microsoft Publisher to start the application.

Whichever route you take to create your new publication, you eventually find yourself in the Publisher application window. For the moment, you might want to close the Catalog by clicking the Close (×) button in the upper right corner of the Catalog window and look at the geography of the Publisher window itself. This window is where your publication appears. Figure 2.2 shows the various parts of the Publisher window, which currently holds a blank publication.

Notice that the largest area of the window is devoted to your publication. All the other areas, such as the menu bar, the toolbars, the Quick Publication Wizard, and the status bar, either provide a fast way to access the various commands and features you use in Publisher or supply you with information concerning your publication, such as what page you are on and where the mouse pointer is currently located on the publication page.

Getting Started with Publisher 13

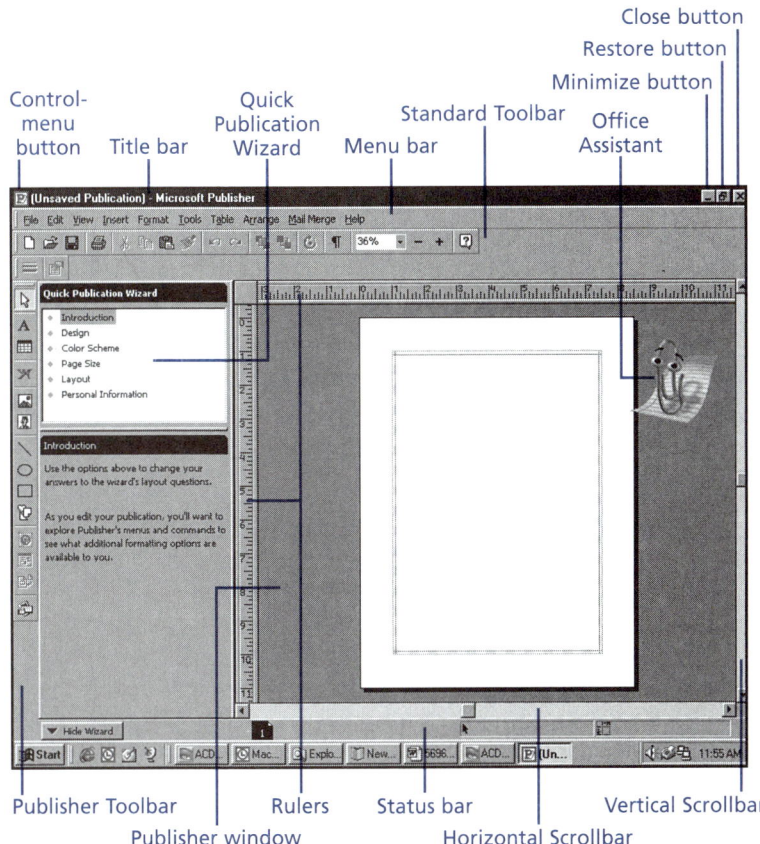

FIGURE 2.2 The Publisher window is where you create your publications and access Publisher's tools and commands.

Table 2.1 describes the elements you see in the Publisher application window.

TABLE 2.1 ELEMENTS OF THE PUBLISHER WINDOW

Element	Description
Title bar	Includes the names of the application and current document, plus the Minimize, Maximize, and Close buttons.

continues

TABLE 2.1 CONTINUED

Element	Description
Control-menu button	Opens the Control menu, which provides such commands as Move, Size, Minimize, and Close.
Minimize button	Reduces the Publisher window to a button on the taskbar; to restore the window to its original size, click the button on the taskbar.
Restore/Maximize button	Enlarges the Publisher window to cover the Windows desktop. When the window is maximized, the Maximize button changes to a Restore button you can click to return the window to its previous size.
Close (×) button	Closes the Publisher program.
Menu bar	Contains menus of commands you can use to perform tasks in the program.
Standard toolbar	Includes icons that serve as shortcuts for common commands, such as Save, Print, and Spelling.
Publisher toolbar	Contains icons for tools that can be used to draw objects and add pictures and text frames to your publication.
Status bar	Displays information about the current page number and the mouse pointer's position on the horizontal and vertical rulers. Information is also provided concerning the height and width of an object you draw on the publication page.
Publication window	Provides you with a place to create and enhance your publication.

Element	Description
Scrollbars	Enable you to scroll the view of your current document: left and right with the horizontal scrollbar and up and down with the vertical scrollbar.
Quick Publication Wizard	Makes it easy for you to change major properties of the current publication, such as the design, color scheme, or page size and layout.
Office Assistant	Provides you with the help you need to get the most out of Publisher. Click the Office Assistant when you need help.

Deciding How to Create a New Publication

As already mentioned, when you first open the Publisher window, the Publisher Catalog appears (see Figure 2.3). The Catalog window provides you with three tabs that supply you with three different ways to create a new publication:

- **Publications by Wizard**   The Publication Wizard provides you with a step-by-step process to creating a new publication. The wizard asks you questions, and the new publication is built based on your answers. A large number of wizard-based publications are available in a series of publication categories. For instance, when you click Invitations in the Wizards pane of the Catalog window, you are provided with thumbnail previews of several different invitation types. You select the invitation type you want to create and then click the Start Wizard button to begin the publication creation process. Creating a publication using a wizard is discussed in detail in Lesson 3, "Creating a New Publication."

- **Publications by Design**   Design sets enable you to create a family of publications that have the same look. Each master design set uses a particular set of design elements and colors that are consistent across all the publications in the set. For instance, you might want to create letterhead, business cards, and invoices that all have the same design look for your small business. You can create these publications by choosing a design set in the Design Sets pane and then selecting the particular publication (such as the business cards) in the Master Sets window. You then start the publication creation process by clicking the Start Wizard button. Using and understanding design sets is discussed in Lesson 4, "Using Design Sets and Templates."

- **Blank Publications**   A third possibility for creating your new publication is to create your publication from scratch. However, Publisher doesn't totally abandon you when you take this approach. The Blank Publications tab actually provides a number of blank templates that help you with the initial layout and page orientation of the new publication. For instance, if you want to create a side fold card, you can select the side fold card template on the Blank Publications tab. After you select a template, you then click the Create button. The techniques used to create blank publications are discussed in many of the lessons in this book. These techniques also apply to editing existing publications.

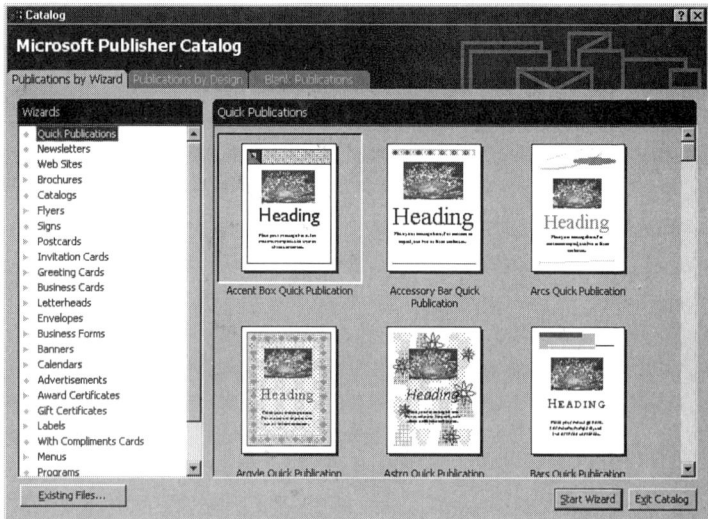

FIGURE 2.3 The Publisher Catalog window provides you with three different ways to create your new publication.

> **Use the File Menu to Open the Catalog** If you inadvertently close the Catalog, select the File menu, and then select New to reopen it.

As far as selecting a route for creating your new publication, that depends on your experience with Publisher and the design aspects of your publication. The Publication Wizard and the design sets provide you with a lot of help as you initially design your publication; in both cases this help is provided by a wizard. Wizard-based publications not only help you select the layout and color aspects of your publication, but they also create placeholder objects in your new publication that you can replace with your own pictures or design elements

For the new user, the wizards offer a quick and easy way to create a new publication. However, if you want to create a publication where you select all the aspects of the publication design, creating a publication from scratch might be easier than creating a publication using a wizard and then modifying that wizard-based publication to meet your needs. However, creating publications from scratch might not be very useful until you have a good understanding of all the Publisher tools and some basic design principles.

How you create your publication also relates to the plan that you have formulated regarding a particular publication or set of publications. For some tips on planning your publications, see Lesson 3.

USING MENUS AND TOOLBARS

After you begin a publication, either from scratch or by using the Publication Wizard or the design sets, you work with the various commands and features that enable you to edit and enhance the publication. Publisher provides you with several different ways to access the commands and features that you use as you work on your publications. You can access these commands using the menus on the menu bar and the buttons on the toolbars.

You can also access a number of Publisher commands using shortcut menus. These menus are accessed by right-clicking on document elements. The shortcut menu appears with a list of commands related to the item that you are current working on, such as a word or paragraph.

THE PUBLISHER MENU BAR

The Publisher menu bar gives you access to all the commands and features that Publisher provides. Like all Windows applications, Publisher's menus are found below the title bar and are activated by clicking on a particular menu choice. The menu opens, providing you with a set of command choices.

Publisher 2000, like the other Office 2000 applications, has adopted a new menu system that provides you the capability to quickly access the commands you use most often. When you first choose a menu, you find a short list of menu commands with the most recently used commands listed first. As you use commands, Publisher adds them to the menu list. To access a menu, follow these steps:

1. Select the menu by clicking its title, such as Insert, as shown in Figure 2.4. The most recently used commands appear; wait a moment for all the commands on a menu to appear.

2. Select the command on the menu that invokes the particular feature, such as Object, as shown in Figure 2.4.

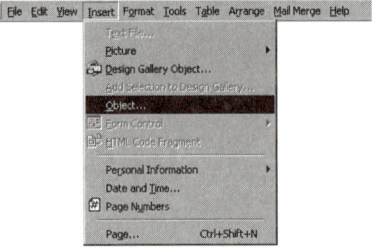

FIGURE 2.4 Select a menu to view, and then select a Publisher command from the menu.

A number of the commands found on the menu are followed by an ellipsis (…). These commands, when selected, open a dialog box that requires you to provide Publisher with additional information before the feature or command can be used. More information about working with dialog boxes is found later in this lesson.

Full Menus at Once If you want to view all the commands available for the menu system, click the Tools menu, click Options, and then click the General tab on the Options dialog box. To show all the commands on the menus, clear the **Menus show recently used commands first** check box. This book covers the various features of Publisher with the personalized menus turned off.

Some of the menus also contain a submenu, or cascading menu, that you can use to make your choices. The menu commands that produce a submenu are indicated by an arrow to the right of the menu choice. In cases where a submenu is present, you point at the command (marked with the arrow) on the main menu to open the submenu.

The menu system provides you with a logical grouping of the Publisher commands and features. For instance, commands related to files such as opening, saving, and printing are all found on the File menu.

Activating Menus with the Keyboard You can also activate a menu by holding down the **Alt** key and then pressing the keyboard key that matches the underscored letter in the menu's name. This underscored letter is called the hotkey. For instance, to activate the File menu in Publisher, you press **Alt + F**.

SHORTCUT MENUS

A fast way to access commands that are related to a document element is to hover the mouse over that publication object and then right-click. This opens a shortcut menu that contains commands related to the object you are working with.

For instance, if you select an object such as clip art or a drawn object such as a circle, right-clicking on the selected object (see Figure 2.5) opens a shortcut menu with commands such as Cut, Copy, and Paste, and other commands related to that particular object.

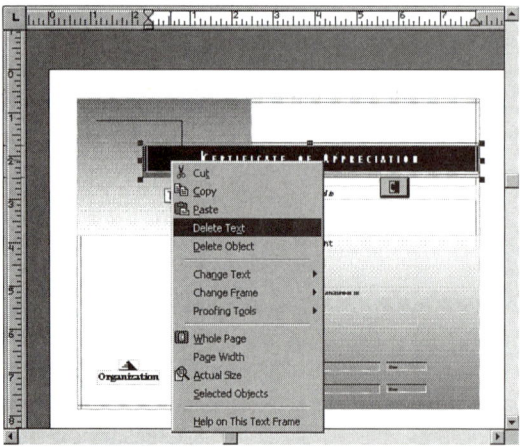

FIGURE 2.5 Right-click to quickly access Publisher commands using shortcut menus.

 Publisher Object An object is any element found in a publication, such as text, a graphic, a hyperlink, or other inserted item. Objects reside in frames that are discussed in Lesson 8, "Working with Publication Frames."

PUBLISHER TOOLBARS

The Publisher toolbars provide you with a very quick and straightforward way of accessing commands and features. When you start a new publication, you are provided with the standard, formatting, and Publisher toolbars. The standard and formatting toolbars reside under the Publisher menu bar. These toolbars provide you with easy access to commands such as Save and Open or formatting features such as Bold, Underline, and Centering, respectively.

To access a command using a standard or formatting toolbar button, click the button. .Depending on the command, either you see an immediate result in your publication (such as the bolding or centering of a selected text object) or a dialog box appears and requests additional information from you.

Where Is the Formatting Toolbar? In Publisher, the formatting toolbar only appears when you have an object in your publication selected. The actual formatting buttons available on the toolbar depend on the type of object selected. For more about objects, see Lessons 10–13.

The Publisher toolbar, which provides quick access to drawing tools and other object-related commands, resides along the left edge of the Publication window. Many of the tools on this toolbar are covered in Lesson 13, "Working with Line Spacing, Tabs, and Numbered Lists."

Finding a Toolbar Button's Purpose You can place the mouse pointer on any toolbar button to view a description of that tool's function.

The Publisher toolbars provide you with a quick fix whenever you need to fire off a command. .Buttons exist for all the commands that are available on the Publisher menu system.

Understanding Dialog Boxes

When you are working with the various commands and features found in Publisher, you invariably come up against dialog boxes. Dialog boxes are used when Publisher needs more information from you before it can complete a command or take advantage of a special feature. Dialog boxes always appear when you select a menu command that is followed by an ellipsis. Dialog boxes also appear when you invoke the same commands using the appropriate toolbar buttons.

The way you provide information in the dialog box can vary. Some dialog boxes need you to type a text entry in a text box. Other dialog boxes provide you with a list of choices on a drop-down menu, and others need you to make your selection by clicking an option button.

Figure 2.6 shows the Font dialog box. This dialog box enables you to make selections using check boxes, drop-down lists, and radio buttons. Some dialog boxes also contain text boxes, where you set an option by typing a parameter or value into a box.

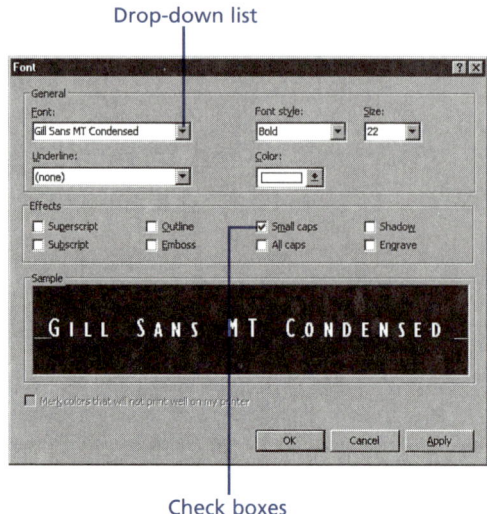

FIGURE 2.6 Dialog boxes ask you for additional information regarding a Publisher command or feature.

Dialog boxes vary, and each might use a different way to get you to supply the needed information for a particular Publisher feature. In most cases, when you complete your selections in a dialog box, you click the OK button to close the box and complete the command that you originally started with a menu choice or toolbar button.

EXITING PUBLISHER

When you complete your initial survey of the Publisher application window or whenever you complete a paricular publication, you might want to exit the software. There is actually more than one way to close the Publisher window, which is the same as exiting the program.

You can exit Publisher by selecting the File menu and then Exit. You can also close Publisher with one click of the mouse by clicking the Publisher Close button (×) in the upper right of the application window.

When you do close Publisher, you might be prompted to save any work that you have done in the application window. If you were experimenting as you read through this lesson, you can click No. The current document is not saved, and the Publisher application window closes. All the ins and outs of actually saving your publications are covered in Lesson 3.

In this lesson, you learned how to start Publisher and explored the various parts of the Publisher window. You also learned how to work with the menu system, toolbars, and dialog boxes. Finally, you learned how to exit the Publisher program. In the next chapter you learn how to create a new document and save your work.

LESSON 3
CREATING A NEW PUBLICATION

In this lesson, you learn to plan and create a new publication using the publication wizards. You also learn how to select publication color schemes and other attributes and how to use personal information sets to automatically place information into your publications.

PLANNING YOUR PUBLICATION

Publisher makes it very easy for you to create professional-looking publications without a massive amount of effort. However, even though Publisher does do a great deal of hand-holding during the publication creation process, you might want to take a little time to plan your publication before you invest the effort to create it.

Planning your publication can really be as simple as asking a few questions that enable you to get a good feel for the type of publication you actually want to create. The list below provides some of the questions you might ask yourself before you actually dive into the publication creation process:

- **What is the purpose of my publication?** Is this publication a greeting card for a friend or business cards that help define your small company to your clients? Obviously, the look and feel of the publication varies a great deal depending on its purpose.

- **Do I have any examples of publications that I really like?** You have probably had the opportunity to examine a brochure or have been given a business card that you really found eye-catching and unique. Adapting layouts and designs you like can help you create great-looking publications. You don't have to reinvent the wheel every time you start a new publication.

- **Is this publication going to be part of a family of publications?** Are you creating the first of many publications for your home business or small company? You might want to take advantage of Publisher's design sets to create an entire family of publications (such as business cards, envelopes, and a letterhead) that have the same look and feel.

- **Does my publication involve special papers?** If you are using special papers to create brochures, business cards, or other publication types, make sure that the color scheme and design elements you choose work well with any pre-existing design elements or colors on the special paper.

- **How can I print the completed publication?** Printing a publication on an inkjet printer that supplies 300 dots per inch gives you decidedly different results than a publication printed by a service bureau or print house (professional printers that print your publication on high-end printing equipment). Gauge your use of colors and overlapping design elements based on the quality you will get during the printing process. If you only have access to a grayscale laser printer, designing a publication extremely rich in color is a waste of your time.

If you at least take the time to consider some of these questions, you might find that your completed publication better suits your means and needs.

USING THE PUBLICATION WIZARDS

An extremely straightforward method of creating a new publication that also provides you with a lot of help is using one of the publication wizards. Taking advantage of a wizard not only ensures that your publication layout and orientation are appropriate for any special papers you are using (such as brochure or business card paper), but the various wizards also help you with the color schemes and design elements in your publication.

When you first start Publisher, the Publisher Catalog appears. You can start your new publication immediately using a wizard by selecting the Publications by Wizard tab.

If you're already working in Publisher, you can open the Publisher Catalog by following these steps:

1. In the Publisher window, click the File menu, and then click New.

2. If you are asked to save changes to any current publication, click Yes to save the changes or No to clear the publication from the Publisher window. The Publisher Catalog appears.

3. Click the Publications by Wizard tab if necessary. The Wizards pane and the publication preview window appear as shown in Figure 3.1.

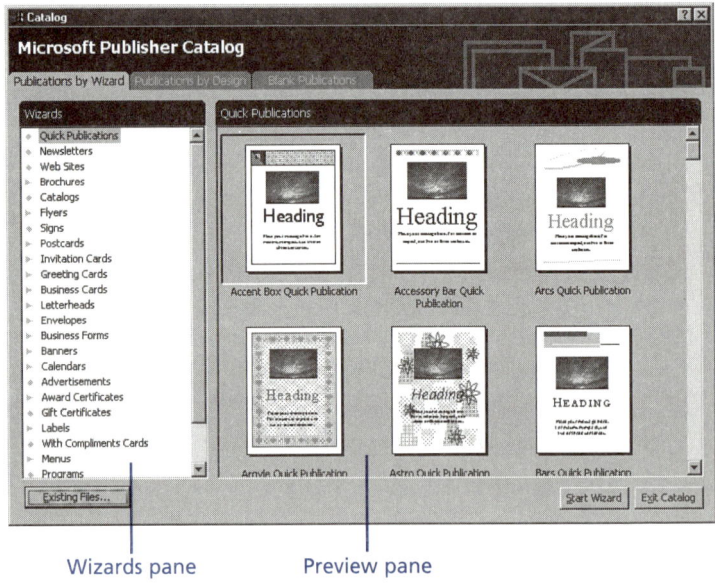

FIGURE 3.1 The Publications by Wizard tab provides you with access to all the publication wizards that Publisher has to offer.

SELECTING A PUBLICATION CATEGORY

When you have access to the Wizards pane, you are provided with a list of different publication types ranging from Quick Publications (which consists of one page flyers) to Origami (yes, Publisher can help you

Creating a New Publication 27

design an Origami boat or crane, as well as others). When you click a particular category, a set of thumbnail previews is provided in the preview window.

It's at this stage of the publication creation process that you select your publication type in the Wizards pane and a layout in the preview pane. After you make your choices, you can actually start the particular wizard that helps you create the chosen publication.

To choose your Publication category and layout, follow these steps:

1. Select a publication category in the Wizards pane.
2. Select a publication in the preview window (see Figure 3.2).

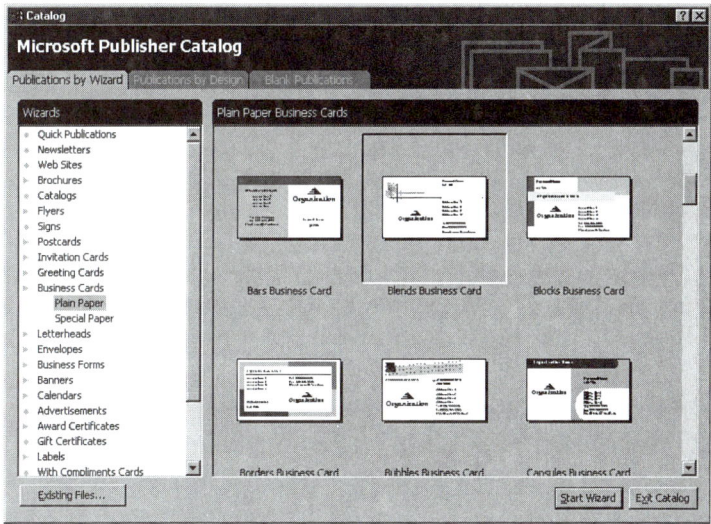

FIGURE 3.2   Select a publication category and then select a specific publication to create.

3. Click the Start Wizard button in the lower right corner of the preview window.

The wizard you start creates the new publication, which appears in the Publication window. The wizard walks you through the publication creation process.

Double-Click to Start Your New Publication You can also select a publication in the preview window and start its associated wizard by double-clicking on the publication preview itself.

SELECTING A PUBLICATION COLOR SCHEME

After you select a particular publication, the next series of steps are controlled by the specific publication wizard. The wizard poses its question in a pane on the left side of the Publisher window. The changes that you make to the publication appear in the pane on the right side of the Publisher application window.

The Wizard Comes First When you choose to use a wizard to create your new publication, if you click anywhere else in the Publisher window (on the menus or the publication itself) you get a message that the wizard is available to help you design your publication. You can quickly bypass the wizard questions, if you so choose, by clicking the Finish button in the wizard pane.

The first time you use one of the publication wizards to create a publication, the publication you create has few or no text entries present in it. This is because information related to you and your company is held in a profile file that is used by the wizard to fill in certain blanks on the publication (such as your name, phone number, and company name). One of the wizard steps is to either create or modify this profile. Supplying profile information is covered in the "Creating a Personal Profile" section later in this lesson.

The first of the publication attributes you are asked to choose is the color scheme for the publication. You have the option of going with the default that was provided when you selected a particular publication for creation, or you can choose your own scheme.

Creating a New Publication 29

To begin using the wizard and select a color scheme, follow these steps:

1. The first wizard screen explains how to use the wizard. When you are ready to begin, click the Next button.

2. The next screen lists the color schemes available for your publication. You can preview a particular color scheme by clicking it. The scheme takes effect on the current publication (see Figure 3.3).

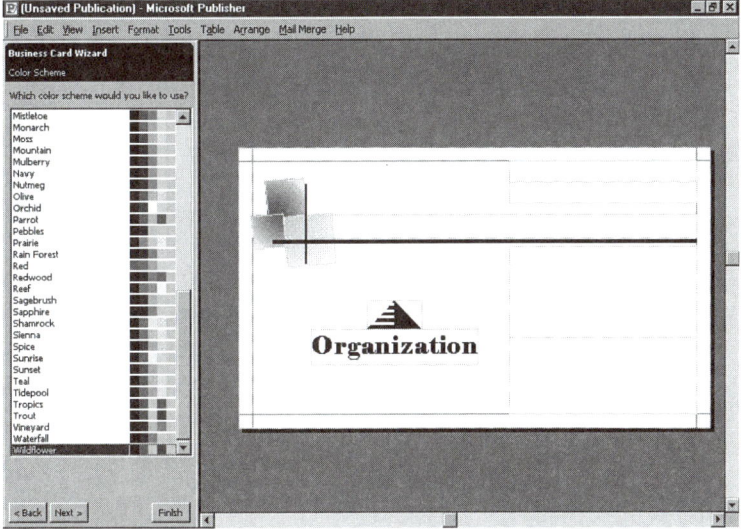

FIGURE 3.3 Glance through the various color schemes until you find the one you want to use for the current publication.

3. When you have decided on a color scheme, make sure that it is selected and click Next to continue.

SELECTING PAGE ORIENTATION

The next step provided by the wizard is where you select the orientation for your new publication. You can choose to create the publication in Portrait (where the publication is orientated on paper from top to bottom, and the height of the paper is greater than the width), or you can select Landscape (where the height of the paper is less than the width), where the paper is rotated ninety degrees.

Portrait or Lansdcape? If you are dealing with a regular 8.5"×11" sheet of paper, Portrait orientation means that the height is 11" and 8.5" is the width. If you use Landscape orientation, the paper is turned on its side and the height is 8.5" and the width is 11".

Depending on the type of publication you are creating, the default page orientation usually provides the best layout for the publication. For instance, business cards are typically printed in a landscape orientation.

To select the orientation for your publication, click either the Portrait or Landscape radio button. When you complete your selection, click the Next button to continue with the publication creation.

Understanding Placeholders

Depending on the type of document you are creating, the next step provided by the wizard might ask if you want to include various placeholders in your publication. These placeholders can be for company logos, pictures, or other graphic elements. For instance, in the case of business cards, the wizard asks you if you want to include a logo placeholder on the business card publication that you are creating. If you have a scanned image or a graphic of your company logo you can replace the placeholder with your image after completing the wizard steps.

When the wizard asks if you want to include a particular type of placeholder in your publication, select Yes to include the placeholder or No to remove the placeholder from the publication. If you are asked what type of content you want in the publication, such as pictures and text, choose from the list provided. Then click Next to continue.

Wizard Steps Can Vary Depending on the type of publication that you are creating, the number of steps provided by the wizard might vary. Read each of the wizard screen's carefully as you work with the various publication types.

CREATING A PERSONAL PROFILE

Depending on the type of publication you are creating, the wizard might prompt you to create a personal profile. This profile contains information like your name, company name, phone number, and other information. In the cases of business forms, business cards, and letterheads, you are prompted to create a new personal profile or edit the existing profile during the wizard-based publication creation. Figure 3.4 shows a Business Card Wizard that has created a blank business card because no information is available in the personal profile.

The best thing about the personal profile is that you enter the information once, and it can be used again and again as you create your various publications.

FIGURE 3.4   Business forms and publications such as business cards derive information automatically from your personal profile.

To create or edit the personal profile, follow these steps:

1. Click the Update button in the Wizard pane. The Personal Information dialog box appears.

2. Use the personal information set list box to choose the type of profile you want to edit or create (Primary Business, Secondary Business, and so on).

3. After you select the information set, fill in the various text boxes (Figure 3.5).

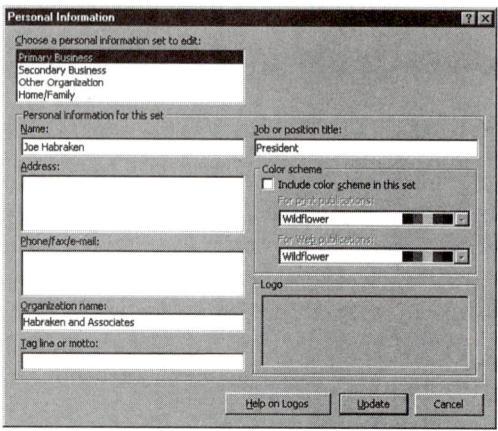

FIGURE 3.5   Fill in the text boxes in the Personal Information dialog box to create a personal information profile.

4. After supplying the information, you can click another profile set in the information set box and fill in the information by following the above steps. When you complete your various information sets, click Update.

The information that you provide in the information set is placed in the publication. Typically, supplying the information profile for the publication is the last step in wizard-based publication creation. You can click Finish to close the wizard and complete the publication.

SAVING YOUR PUBLICATION

After you complete the wizard steps, save your publication. This enables you to take a breather before you begin the editing or enhancement process.

To save your new publication, follow these steps:

1. Select the File menu, and then select Save. The Save As dialog box appears.

2. Type a file name in the file name box.

3. Click the Save in drop-down box, and select the drive you want to save the publication in.

4. Folders in the selected drive appear in the Save As box. Double-click a folder for the publication to reside in.

5. Click the Save button.

The publication is saved to your computer.

In this lesson, you learned how to create a new publication using the publication wizards. You also learned how to select color schemes and create a personal information profile. In the next lesson, you learn how to create publications from scratch and by using design sets.

Lesson 4

Using Design Sets and Templates

In this lesson, you learn to create new publications as part of a design set. You also learn to create publications from scratch.

Understanding the Publication Design Sets

Publisher provides you with a way to create sets of publications that share the same color and design attributes. This makes it easy for you to create small business publications, such as business cards, envelopes, brochures, and letterheads, that share the same look.

The great thing about using the design sets is that not only do you end up with a group of publications that share the same basic design and color elements, but when you create each of the individual publications in the set, you are walked through the creation process by a wizard (in exactly the same manner as when you choose to create individual publications using a wizard as discussed in Lesson 3, "Creating a New Publication").

Selecting the Design Set

You select the design set for your new publication from the Publisher Catalog, which appears when you start Publisher or when you start a new publication from the File menu. To select a design set for a new publication, follow these steps:

1. Start Publisher using the Start menu, or if you are already in the Publisher window, click the File menu and then click New.

2. If you are asked to save changes to any current publication, click Yes to save the changes, or click No to clear the publication from the Publisher window. The Publisher Catalog appears.

3. Click the Publications by Design tab. The Design Sets pane and the Master Sets preview window appear, as shown in Figure 4.1.

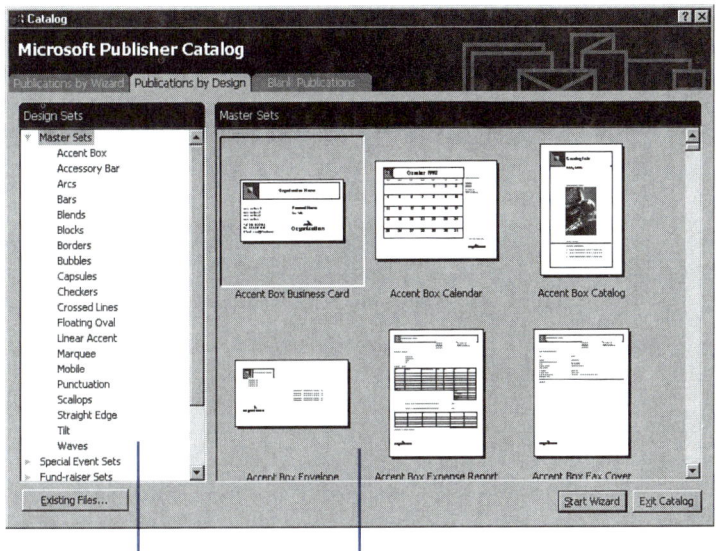

FIGURE 4.1 The Publications by Design tab of the Publisher Catalog enables you to create publications with a unified design.

4. The design sets are listed in the Design Sets pane on the left side of the Publisher window. There is a master design set for business publications, and there are specialty design sets including Special Event Sets and Holiday Sets. To expand one of the set categories, such as the Master Sets or the Special Event Sets, click the set category. The design sets under that particular category appear.

5. To view the publications included in a particular design set, click the set in the Design Set pane. A preview of the publications in the set appears in the Master Sets window (see Figure 4.2).

36 Lesson 4

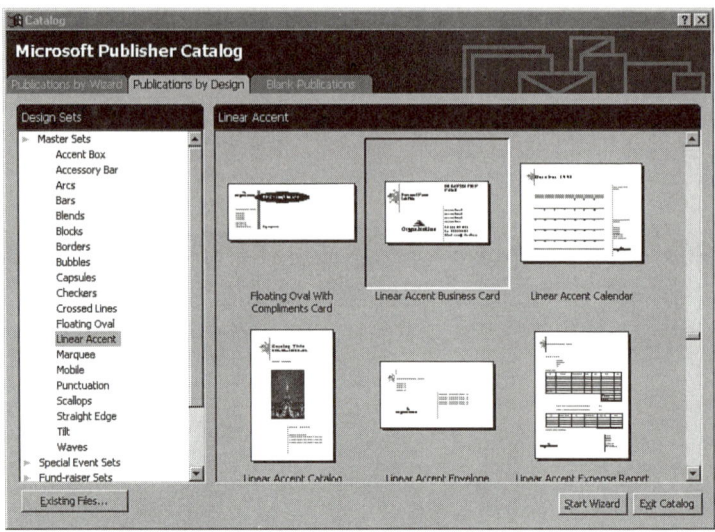

Figure 4.2 Each design set contains several different publication masters that embrace the same design elements.

6. After you preview the various design sets that include the type of publication you want to create (such as a brochure or business cards), select the design set in the design set list.

 Use the Scrollbar to Peruse the Design List To view all the design sets in the Design Sets pane, click the down scroll arrow on the vertical scrollbar.

Selecting a Design Master

After you select a particular design set, you can choose the actual publication to create using the design elements of that particular set. The publications available for the set are previewed in the right pane of the Catalog window. (The title of the pane changes to the design set that you choose in the left pane.)

To select your design master and begin the publication creation process, follow these steps:

1. Click a publication master in the publication pane of the Catalog window (see Figure 4.3).

2. Click the Start Wizard button in the lower right corner of the preview window.

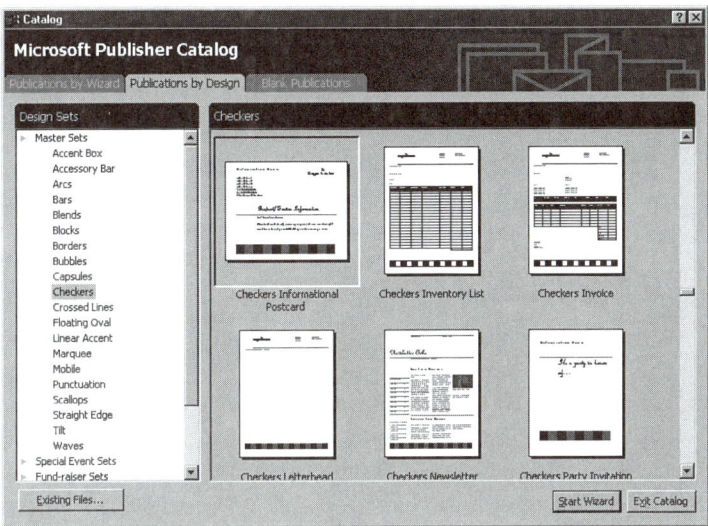

FIGURE 4.3 Select a publication category and then select a specific publication to create.

The wizard creates the new publication, which appears in the Publication window. The wizard walks you through the publication creation process.

COMPLETING THE PUBLICATION

Completing the publication merely requires that you answer each question posed by the specific publication wizard, which appears in the left pane of the Publisher window. Your publication, using the design set you chose on the Design Sets tab, appears in the Publication window.

Complete each wizard step as you would for any wizard-created publication (for more about the publication wizards, see Lesson 3). However, when you are given the option of changing the color scheme for the particular publication, accept the default for this step. Remember that your strategy in using the Design Sets tab is to create a group of publications that embrace the same color scheme and design elements. If you change the color scheme for this particular publication, it no longer matches other publications you create using the design set.

When you complete the various steps provided by the wizard (when you click Finish), make sure to save your new publication. You can quickly open the Save As dialog box and assign a name to the new publication using the Save button on the Publisher Standard toolbar. After you assign a name and a location to the new publication, click the Save button in the Save As dialog box.

ADDING YOUR OWN DESIGN AND COLOR SCHEMES

If you want, you can also create a publication from scratch. This means that all design elements and color schemes have to be added to the publication manually. Publisher does provide you with a set of templates, however, that help set up the page size and orientation for blank publications. You can use templates to create business cards, Web pages, tent cards, and other publication types from scratch.

To start the process of creating a new publication from scratch, follow these steps:

1. With the Publisher Catalog open (the Catalog opens either when you start Publisher or click the File menu, and then click New), click the Blank Publications tab. The Blank Publications pane appears on the left side of the Catalog and the Blank Full Page preview window appears on the right, as shown in Figure 4.4.

2. Select the template you want to use for your blank publication either by clicking a choice in the Blank Publications list (in the left Catalog pane), or by clicking one of the publication previews in the preview window.

3. Click the Create button.

Using Design Sets and Templates 39

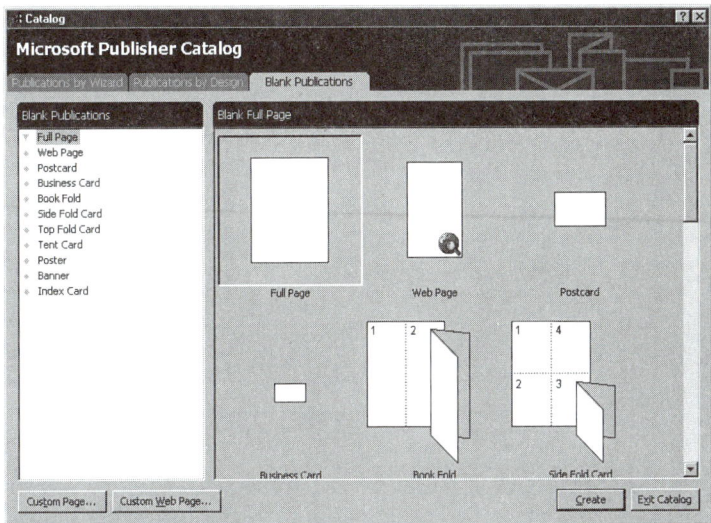

FIGURE 4.4 You can base your blank publication on several templates provided by Publisher.

The blank publication appears in the Publication window. On the left side of the Publisher window, the Quick Publication Wizard pane appears. You can select any heading available in the Quick Publication Wizard to edit the design parameters of your blank publication.

For instance, if you want to assign a color scheme to the blank publication, click the Color Scheme heading. The Color Scheme pane appears (see Figure 4.5). Use the scrollbar to scroll through the Color Scheme list; when you find the color scheme you want to use, click it.

You can also select a design (a set of design elements that are added to your publication), page size (portrait or landscape), and layout for your blank publication. Selecting a layout that includes text and pictures also enables you to have information in your profile automatically placed in the publication. Figure 4.6 shows a blank publication that has been customized with a color scheme, design, and layout.

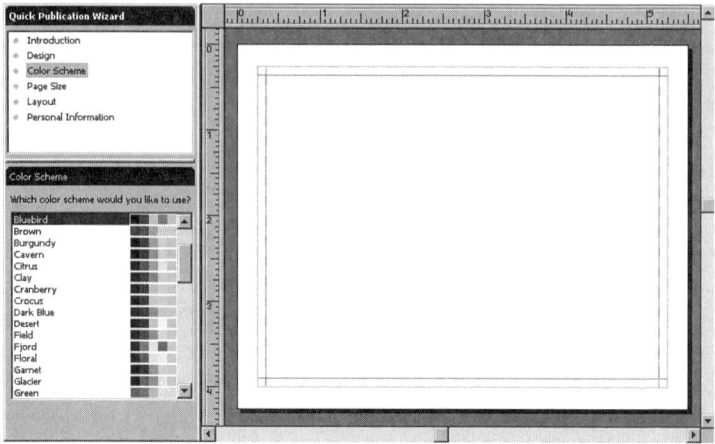

FIGURE 4.5 The Quick Publication Wizard makes it easy for you to assign a color scheme and other design parameters to your blank publication.

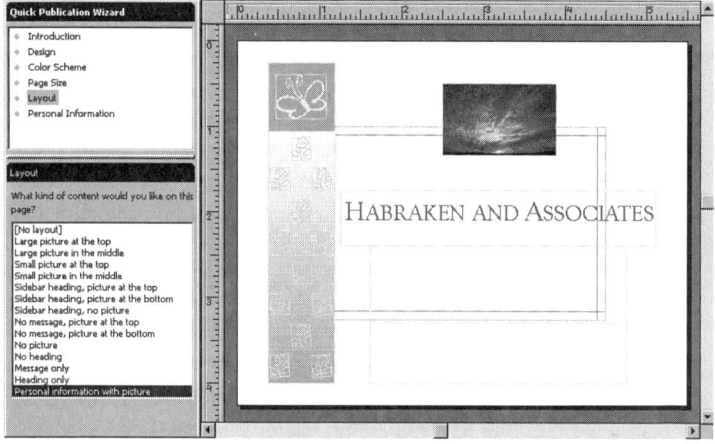

FIGURE 4.6 Blank publications can be enhanced to look as well-designed as those created using a Wizard or one of the design sets.

As you can see, even creating a publication from scratch in Publisher is very straightforward, and Publisher offers a great deal of help in setting the design parameters for the publication. When you complete setting the design parameters for your "from scratch" publication using the Quick Publication Wizard, make sure you save the publication.

You Can Close the Quick Publication Wizard After you set the various design parameters for your blank publication, you can close the Quick Publication Wizard by clicking the Hide Wizard button at the bottom of the wizard pane.

In this lesson, you learned how to create a new publication using the design sets. You also learned how to create a new publication from scratch. In the next lesson you learn how to change the view of your publication and zoom in and out on the pages.

Lesson 5

Viewing Your Publications

In this lesson, you learn the different options for viewing your publication, including the Zoom feature. You also learn to work with positioning tools such as the ruler and guides.

Changing the Publication Display

When you create publications in the Publisher window, the default view is the Whole Page view. This enables you to see the entire current page from a sort of bird's eye view that is excellent for determining the overall layout of the page and the positioning of the various text frames, picture frames, and other objects.

As you work with your publications, it's convenient to be able to change the view from a single page to a Two Page Spread and also have the capability to zoom in and out on a particular page. The following sections describe the three basic views that are available to you.

Full Page

The default view is Whole Page. It shows the entire current page and shows the margins for the page. Figure 5.1 shows a publication in the Whole Page view.

If you are in any other view and want to return to the Whole Page view, select the View menu and then click Zoom. Select Whole Page from the cascading menu.

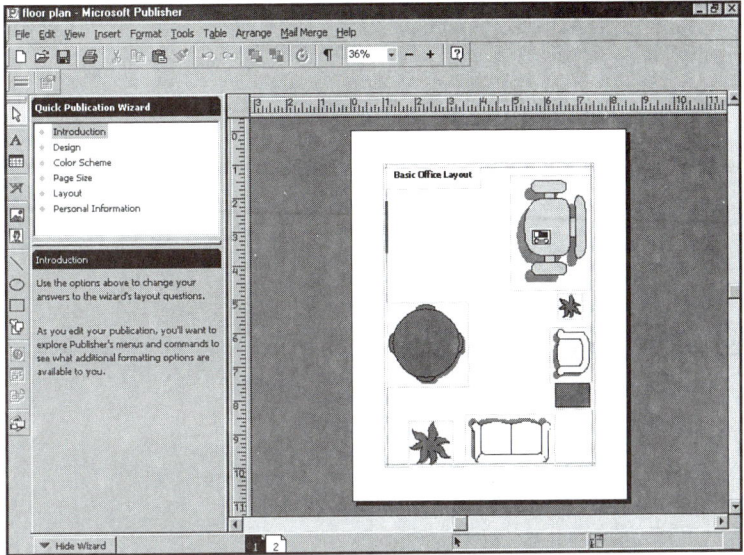

FIGURE 5.1 The Whole Page view enables you to view the spatial relation of all the items you place on a particular page.

 Go to Whole Page Using the Keyboard You can quickly go to the Whole Page view by pressing **Ctrl+Shift+L** on the keyboard.

PAGE WIDTH

Another useful view is the Page Width view. This enables you to zoom in on the publication page but still see the left and right margins. The Page Width view provides a view that is slightly larger than zooming to 50% (58%). Because this view maintains the total width of the page, enabling you to see the left and right margins, you can easily scroll up and down the page using the vertical scrollbar. Figure 5.2 shows the page depicted in Figure 5.1 (the Whole Page view) in the Page Width view.

To place the current publication page in the Page Width view, select the View menu and click Zoom. Select Page Width from the cascading menu.

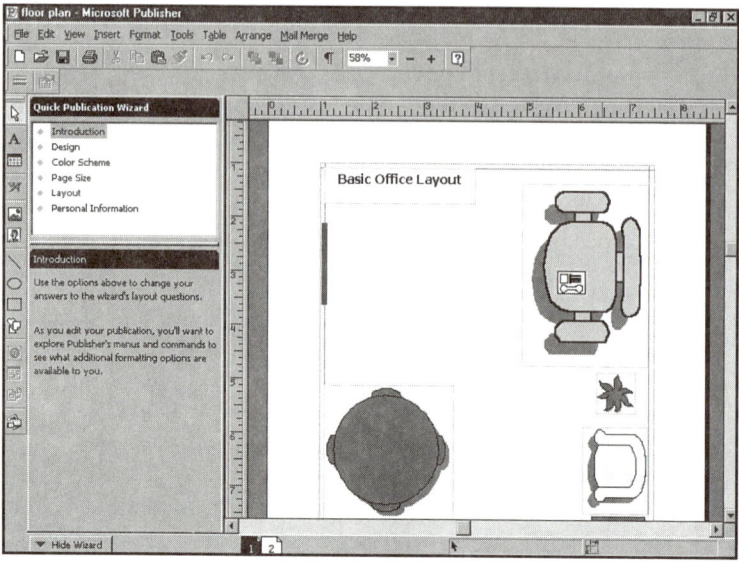

FIGURE 5.2 The Page Width view enables you to zoom in on your publication page but still shows the left and right margins for reference.

Two-Page Spread

Another useful view is the Two-Page Spread. This enables you to examine facing pages in a publication. This is particularly useful when you want to make sure that the frames and objects on these two pages are balanced and arranged appropriately. Any publication that opens has facing pages inside it (for example, a greeting card or brochure).

Figure 5.3 shows a three-page publication with the facing pages (pages 2 and 3) in the Two-Page Spread view. To view a publication in the Two-Page Spread view, select the View menu and then select Two-Page Spread. A check mark appears next to the Two-Page Spread selection on the View menu.

When you want to return to the single page view, select the View menu and click Two-Page Spread to remove the checkmark.

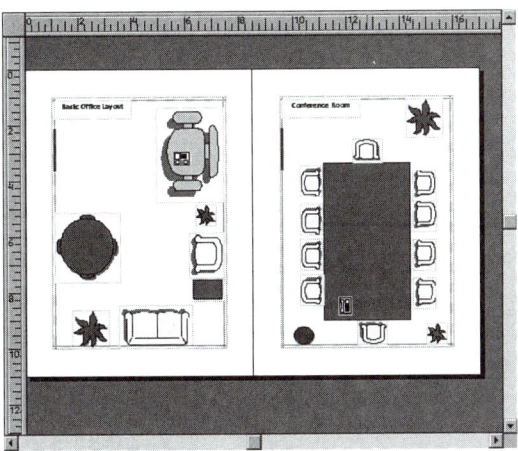

FIGURE 5.3 The Two-Page Spread view enables you to look at facing pages in your publication.

 First Pages Don't Have Facing Pages The very first page of your publication is considered the cover page. If you think about greeting cards and other booklet-type publications, the first page is seen when the publication is closed. The first inside page (on the left of the open card) is page 2, and page 3 is its right facing page.

USING THE ZOOM FEATURE

Because your publication pages consist of various frames containing text and other items such as pictures, you often need to edit or otherwise fine-tune these items. Obviously, when you try to concentrate on a particular item on the page, you want to be able to zoom in on that item. Publisher provides you with the capability to zoom in and out on your publication pages using a range from 10% to 400% (the larger the percentage, the more you're zoomed in on your publication).

To zoom in or out on the current page, follow these steps:

1. Select the View menu and then click Zoom.

2. Select your Zoom percentage from the cascading menu (such as 75%). Figure 5.4 shows a page zoomed at 75%.

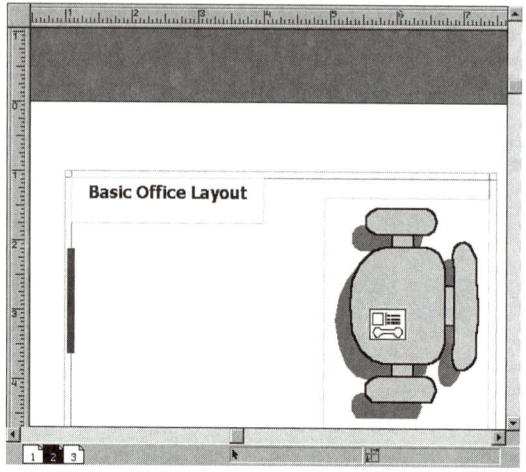

FIGURE 5.4   Use the Zoom feature to zoom in and out on your publication pages.

 Zoom In and Out Quickly Using the Toolbar   You can also zoom in and out on your publication pages using the Zoom drop-down box on the standard toolbar. Click it and select the zoom percentage.

SCROLLING IN THE PUBLICATION

When you zoom in on your publication, you usually need to adjust your point of view up or down or to the left or right to see a particular part of the page or focus on a particular item. The vertical and horizontal scrollbars provide you with the capability to do this.

To move up or down on the publication page, click the up or down arrow on the vertical scrollbar, respectively. To scroll to the right or left, click the right or left scroll arrow on the horizontal scrollbar.

You can also use the scroll boxes on either of the scrollbars to scroll to a particular position in the document. For instance, if you want to scroll halfway down a page, you can drag the vertical scroll box to the middle of the vertical scrollbar (see Figure 5.5).

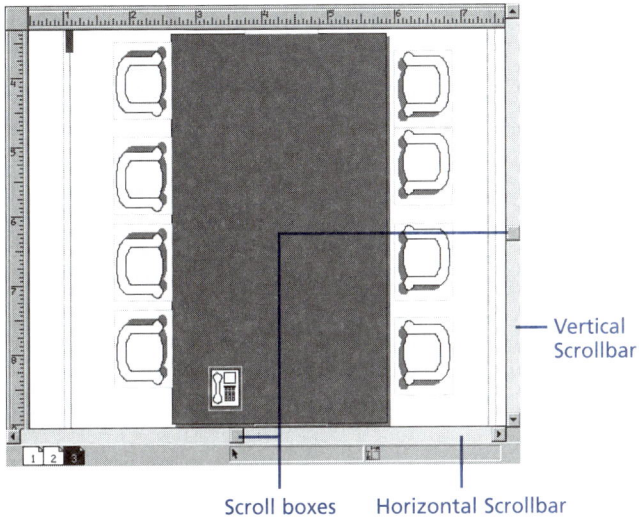

FIGURE 5.5   The scrollbars enable you to scroll vertically and horizontally through a document page.

WORKING WITH RULERS AND GUIDE LINES

Another visual aspect of working on your publications is the use of the rulers and guides. A vertical and a horizontal ruler are supplied in the Publication window to help you place items on the page. *Guides* are really extensions of the rulers and appear as lines (both vertical and horizontal) that you can drag onto the document page from either ruler to help you appropriately place text and pictures on the page.

Guides A layout guide is a nonprinting vertical or horizontal line that you place on the publication page to help you align the various elements the publication contains. Guides appear on the document page in green.

 Where Are My Rulers? If you don't see the rulers in the publication window, select the View menu and then select Rulers.

USING THE RULER

When you move the mouse on the page, the vertical and horizontal positions of the mouse pointer are tracked by a tick mark (a line) on each of the rulers. The actual horizontal and vertical position of the mouse pointer is represented by two intersecting lines called *crosshairs* and is also displayed on the Publisher taskbar (see Figure 5.6).

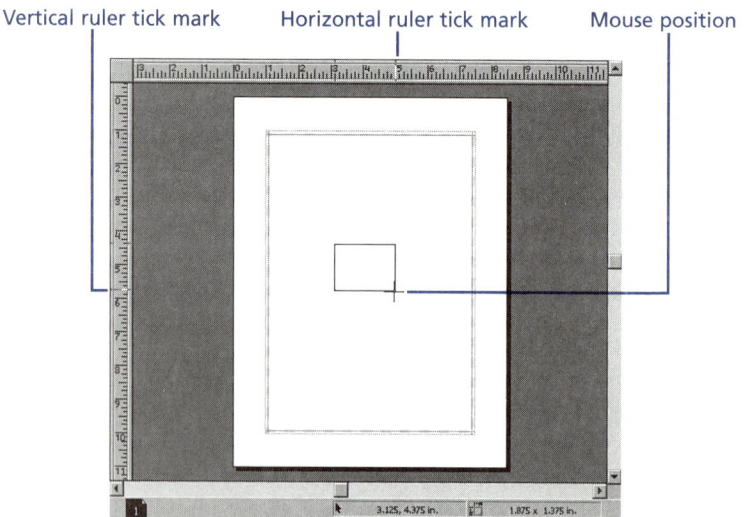

FIGURE 5.6 The rulers show you the position of the mouse pointer on the page.

Being able to see the position of the mouse pointer on the rulers and in the taskbar enables you to position object frames on the page with precision. For instance, to place a picture frame two inches down from the top of a page, you drag the object by its top (because you want the top at the two inch mark) and watch until the vertical ruler tick mark reaches the two inch mark. Then you release the object. For more information about inserting and moving frames on a publication page, see Lesson 8, "Working with Publication Frames."

Using Layout Guides

Another tool you can use to precisely position frames and other items on your Publisher page is the layout guide. These guides, as previously mentioned, are nonprinting vertical and horizontal lines that you place on a publication page. The great thing about using layout guides is that they help you maintain the overall layout design on publications that run a number of pages (you can set up the same layout guides on each page).

Placing a guide on the page enables you to actually place a frame on the guide. This is called a *snap to* guide. When you move the frame near the guide, the frame "snaps" onto the guide, positioning the item.

Turning On Snap to Guides If you are going to use guides in your publications to help you place items on the page, you need to turn on the Snap to Guide feature. Select the Tools menu and then select Snap to Guides.

Guides are created and positioned on your pages using the mouse. Follow these steps to create guides on a publication page:

1. To create a vertical layout guide, place the mouse pointer on the vertical ruler.
2. Hold down the **Shift** key and drag the mouse pointer onto the page. A vertical guide appears (see Figure 5.7).

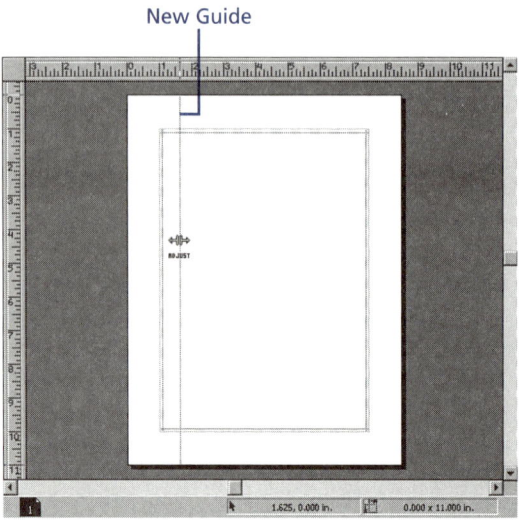

Figure 5.7 Position guides on your pages to help you align text and picture frames.

3. Place the guide at the appropriate position on the page and release the mouse button.

4. To create a horizontal layout guide, repeats steps 1–3, but drag a guide down from the horizontal ruler.

 Where Are My Guides? If you attempt to create guides and they don't appear in the publication window, select the View menu and then select Show Boundaries and Guides.

You can place as many guides as you need on your pages. If you find that you like using guides but don't like creating and positioning them with the mouse, you can also choose to have a series of horizontal and vertical guides created for you automatically. This forms a grid pattern on the publication, providing you with a sort of "topography" that you can use to appropriately align your publication items.

To create a grid system for a page, follow these steps:

1. Select the Arrange menu and then select Layout Guides. The Layout Guides dialog box appears (see Figure 5.8).

2. To set guides, use the Columns and Rows click boxes in the Grid Guides area of the dialog box to set the number of vertical (column) and horizontal (row) guides you want to place on the page.

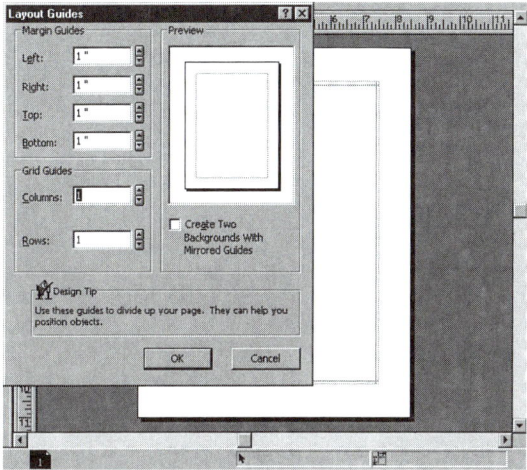

FIGURE 5.8 You can quickly set up an entire grid of guides in the Layout Guides dialog box.

3. After you specify the number of columns and rows you want for the guides, click OK. The new guides appear on your publication page.

 Set Margins Here, Too The Layout Guides dialog box also enables you to set the margins for a particular publication page. Setting publication margins is covered in Lesson 16, "Formatting Publication Pages."

In this lesson, you learned how to change the view of your publication and use the Zoom feature. You also learned how to use the ruler and place guide lines on a publication page. In the next lesson, you learn how to work with and enhance existing publications.

Lesson 6

Working with Existing Publications

In this lesson, you learn to open, close, and save an existing publication; add pages to the publication; and save the publication under a new file name. You also learn to complete wizard-based publications including text frames and placeholders.

Opening an Existing Publication

You will probably end up with a library of saved publications that you use on a fairly regular basis. Items such as certificates, invitation cards, and various business forms can be created and saved to your computer and then used when needed.

 Saving Is Not Just for Finished Publications If you work on a publication and don't really have its design or colors the way that you want them, you can, of course, save the file and then work on it again at your earliest convenience.

The great thing about recycling publications in this way is that you take the time to design them well once and then you can open them and edit them to fit your particular needs. To open an existing publication, follow these steps:

1. In the Publisher window, click the File menu and then click Open. The Open Publication dialog box appears (see Figure 6.1).

Use the Open Button To quickly open a publication, click the Open button on the Publisher standard toolbar.

2. In the Open Publication dialog box, click the Look In drop-down box to select the drive that your file is located on.

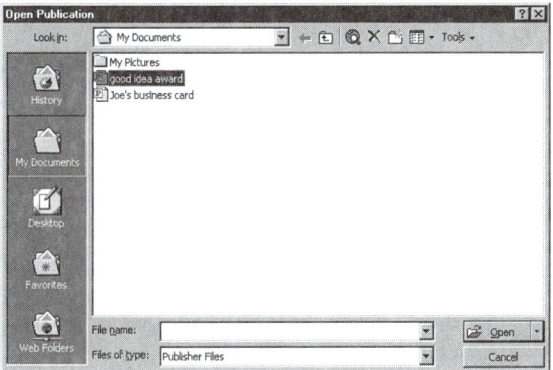

Figure 6.1 Select a location on your computer to open a specific file.

3. After you select a drive, double-click the appropriate folder in the list that appears.

4. Select the file you want to open, and click Open. The publication opens in the Publisher window.

 Saving the Previous Publication Publisher only enables you to work on one publication at a time. If you try to open or start a different publication and have not saved the current publication, you are asked to save the current publication when you click the Open button in the Open Publication dialog box.

Completing a Wizard-Based Publication

When you create publications using a wizard, it's impossible for the wizard to fill in all the information needed in the publication. Of course, a great deal of information is added automatically from the personal information set that you use for the publication. However, in a publication like a certificate or an invitation, you still have to fill in the recipient's name or the time and place of the event respectively. So, after opening a saved publication, you have to edit the content to meet your current needs.

Editing Text in a Publication

When you work with text in Publisher, the text is held inside a frame called a *text frame*. You learn more about creating and working with text frames in Lesson 10, "Changing How Text Looks."

> **Frame** A frame is the container that is created to hold a picture, text, or other object in a publication. A frame by default does not have a border around it that can print, but it can be customized with borders, colors, and other attributes. See Lesson 8, "Working with Publication Frames."

To modify or complete text entries in a publication, follow these steps:

1. To modify the contents of a text frame, click the text frame to select it (click anywhere on the frame around the text).

> **Zoom in When Working on Text** When you need to edit text in a text frame, you might want to zoom in on the publication for a closer look. Click the View menu, click Zoom, and then select the zoom percentage from the cascading menu.

2. Double-click the text in the text frame to select all the text (see Figure 6.2). To select a portion of the text, click at one end of it and drag across it.

3. Type the appropriate text in the text frame. The selected text is deleted and replaced by the new text.

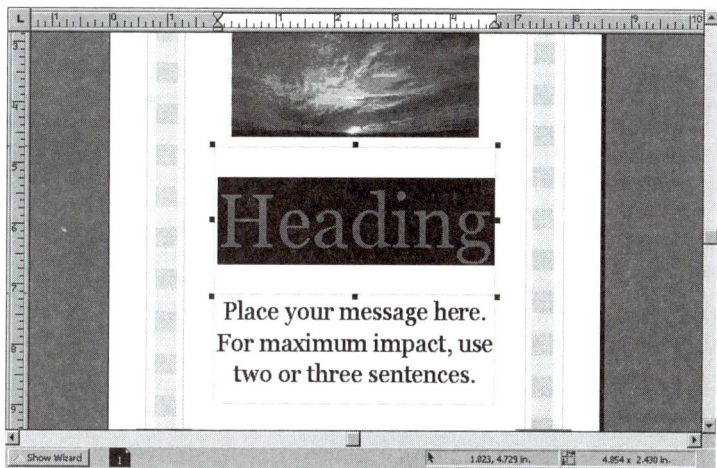

Figure 6.2 Select the text that you want to modify.

FILLING A PICTURE PLACEHOLDER

Not all the additions that you need to make to a wizard-based publication, however, are text. In some cases you need to fill placeholders (such as for a company logo or a picture of the chairperson of the board) with the appropriate graphics. These graphics are held in a frame just like text and are often referred to as *picture frames*. See Lesson 11, "Working with Graphics," for more information on working with picture frames.

To fill a picture placeholder with a new picture, follow these steps:

1. Click the placeholder picture frame in the publication (see Figure 6.3) to select it (it can be a picture frame or some other type of object frame).

Figure 6.3 Select any picture placeholder and insert a picture in its place.

2. Select the Insert menu, and then click Picture. From the cascading menu, select any of the following:

- **Clip Art** You can select a new picture from the Clip Art Catalog.

- **From File** You can select a picture that you have stored on your computer.

- **From Scanner or Camera** If you have either of these devices hooked to your computer, you can create a new image and immediately place it in Publisher.

- **New Drawing** You can draw your own image to replace the placeholder item.

Depending on your preceding choice, you work with the Clip Art Catalog, the Insert File dialog box, your scanner or camera, or the Publisher Drawing tool, respectively. Figure 6.4 shows a picture being selected for insertion into a publication using the From File option. For more information on working with and manipulating pictures, see Lesson 11.

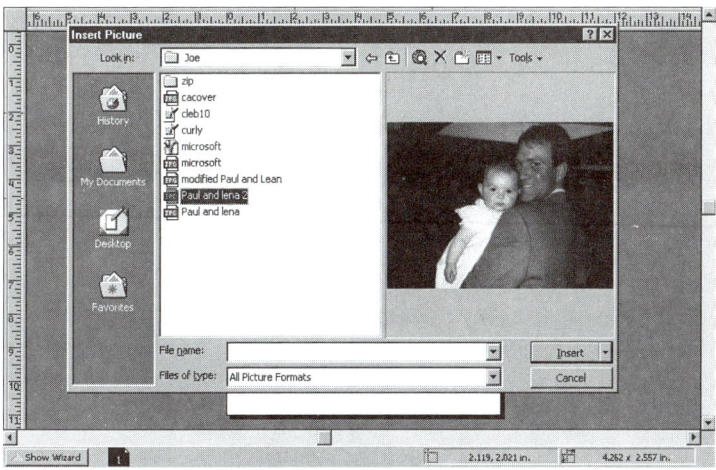

Figure 6.4 You can replace placeholders with Publisher clip art or other pictures that you have stored on your computer.

After you replace a placeholder item with a picture or graphics of your own, you can continue to edit the text and graphics items in the publication. The number of text frames and placeholder items that you need to edit to complete a particular publication depends on the publication itself. The more complex the publication, typically the more items you have to personalize.

ADDING PAGES TO A PUBLICATION

You might find after you complete a particular publication that you need to add additional pages to it. This can be a common need when you build publications from scratch.

To add pages to the current publication, follow these steps:

1. Select the Insert menu, and then select Page. The Insert Page dialog box appears (see Figure 6.5).

2. Type the number of new pages you want to insert in the Number of new pages box.

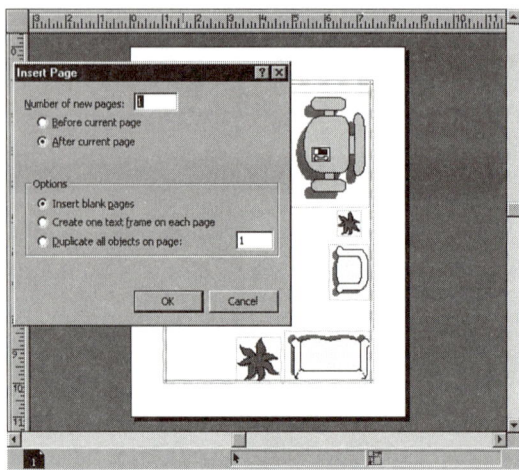

Figure 6.5 You can insert new pages into a publication before or after the current page.

3. Click either the Before current page or After current page radio button to select the appropriate option.

4. In the Objects box select the appropriate option for your blank pages:

> **Insert blank pages**   The pages are inserted into the publication with no frames.
>
> **Create one text frame on each page**   A text frame is placed on each of the new pages.
>
> **Duplicate all objects on page**   This option copies the objects on the designated page (type the number in the page box) and places them on the new page or pages inserted.

After you make your selections, click OK to insert the new page or pages into your document.

Save Your Work When you make changes to your publications, such as adding pages, remember to save your work. Publisher prompts you every so often to save your work if you're remiss. Click Yes in the dialog box that appears to save your publication changes.

SAVING A REVISED DOCUMENT UNDER A NEW NAME

When you work with publications that you use again and again, such as award certificates or invitations, you might want to keep your original publication incomplete (with items left blank until the publication is made ready for printing) under a particular file name. For instance, you might create an award certificate and leave the recipient's name blank. Then when you are ready to create an award for a particular person, you open the award publication and edit the text frame that contains the recipient's name.

You might also want to save the completed publication (the one with the recipient's name or other information) under a different file name. This can be done using the Save As command.

To save a publication under a different file name, follow these steps:

1. Select the File menu, and then select Save As. The Save As dialog box appears.
2. Type a new name for the publication in the File name box.
3. Use the Save in drop-down box to designate the drive that you will save the file to. Also make sure to select a folder on that drive by double-clicking the folder.
4. Click Save to save the file.

You have now saved any changes that you made to your original publication under a new file name. This means that the original file under the original file name still exists and can be used whenever you need it to create personalized publications that can then be saved under a different file name.

Closing a Publication

When you save a particular publication and are finished working with it, you usually close the publication. To close the current publication, click the File menu, and then click Close. The publication closes, and Publisher opens a new blank publication in the Publisher window.

In this lesson, you learned how to open an existing publication, modify the publication, and use the Save As command to change the name of the publication. In the next lesson, you learn how to get help in Publisher.

Lesson 7
Getting Help in Publisher

In this lesson, you learn to use the Office Assistant, to search for help on specific topics, and to use other Help features such as online help.

Using the Office Assistant

If you've already tried to get help in Publisher, you've probably noticed the Office Assistant: He's the helpful paper clip, Clipit, or other special character who sits in the Publisher window waiting to provide you with help. You can use the Office Assistant to help you with procedures, explanations, and tasks you perform in Publisher.

To use the Office Assistant, do one of the following:

- If the Assistant is already in the Publisher window, click the Assistant.
- Choose Help, Microsoft Publisher Help.
- Press **F1**.

The Assistant displays the text balloon shown in Figure 7.1 when you click it. The Office Assistant's help is context-sensitive; thus, depending on where you are in the program, the Assistant tries to offer help on related topics. If you need help on one of the suggested topics, click one of the blue option buttons.

 Context-Sensitivity This is a help feature that senses where you are in the program (working on objects or formatting text) and offers help on topics that are related to your current task.

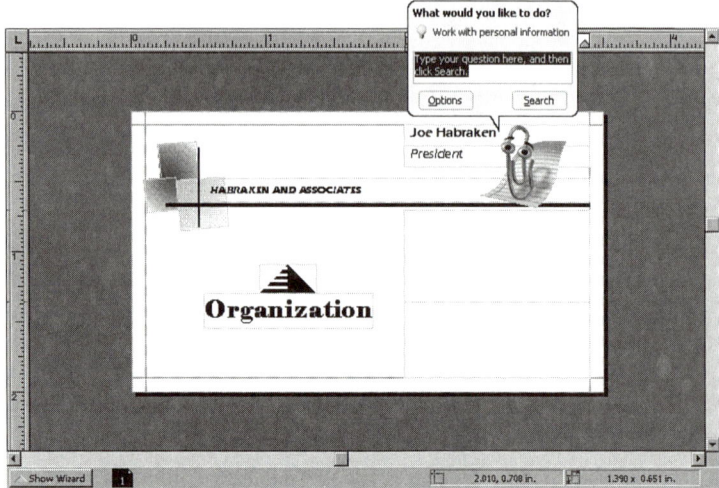

Figure 7.1 Use the Office Assistant to get context-sensitive help in Publisher.

If you don't get a list of topics from the Assistant, you can enter a question in the Assistant's balloon in the search box and click Search. You might, for example, enter "How do I print?" (or "How do I save a file?" or any other question containing key words related to topics for which you need help). When the Office Assistant box appears, it contains options related to your question.

To close the Office Assistant balloon, click outside it.

> **Tricks of the Trade** When a light bulb appears over the Assistant, this means the Assistant has help to offer you on your current task. Click the Assistant to see what it has to offer.

You can right-click the Office Assistant and choose Hide Assistant from the shortcut menu. You can also choose options for the Assistant or choose to use a different character as your Assistant.

To choose a different character as your Assistant, follow these steps:

1. Right-click the current Assistant, and select Choose Assistant from the shortcut menu. An Office Assistant Gallery box appears.

2. Use the Next or Back buttons to view the Assistant characters available.

3. After selecting the new Assistant character, click the OK button. The new Assistant appears in the Publisher window.

 Find that Office CD-ROM You might have to insert your Office or Publisher CD-ROM into your CD-ROM drive to select a new Assistant.

Getting Help Without the Assistant

If you prefer, you can turn off the Assistant feature and access Publisher Help in a more conventional manner. Click the Assistant, and on the Assistant balloon click Options. On the Options tab deselect the check box that says Use the Office Assistant. The Assistant is removed from the Publisher window. Now you're ready to get help from Publisher without going through the Assistant. Click the Help menu; then click Microsoft Publisher Help. The Help window appears and explains different ways to get help in Publisher.

Click the Show button at the top left of the Help window. The window expands and three tabs appear—Contents, Answer Wizard, and Index.

The Help window has the look and feel of a World Wide Web browser. When you look up a particular topic, additional information can be accessed by clicking on a highlighted item that operates as a hyperlink. This link takes you to a new set of help information. The Help window has Back and Forward buttons that enable you to move backward and forward through the Help information that you view (Figure 7.2).

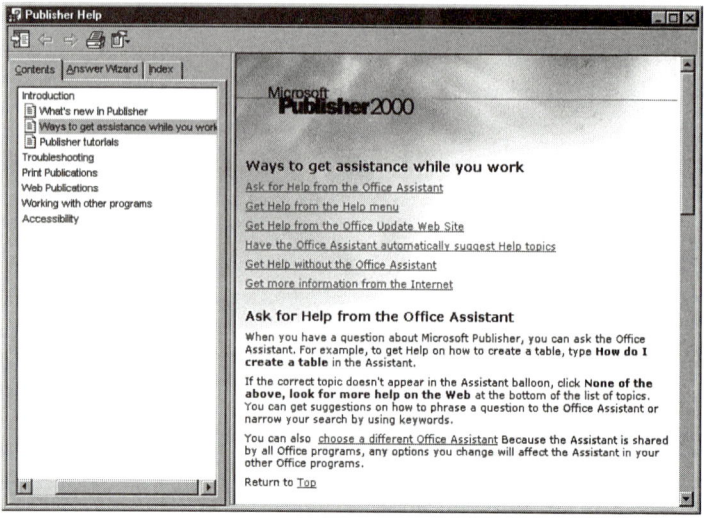

FIGURE 7.2 When you turn off the Assistant, you can directly access the Publisher Help system.

USING THE CONTENTS TAB

Each of the Help tabs provides you with a slightly different way to access the help you need. The Contents tab supplies you with a list that takes you to major groupings of information, such as What's New in Publisher, and information on troubleshooting or printing your publications.

For an example of how the Contents tab works, take a look at how it supplies you with information on troubleshooting problems with objects in your publications. Each main topic on the Contents tab is represented by a book icon. Double-click the Troubleshooting book. The book icon opens up and displays a list of more specific topics. Double-click Objects in the list.

In the right pane of the Help window, you are provided with a list of more specific topics on troubleshooting problems with objects. Click one of these items (such as adding objects or pictures) to view a more definitive list of help topics in the right pane. A second, less general list of help topics related to troubleshooting objects (such as having problems with clip art) appears. Click one of these more specific listings.

Depending on the help item you click in the right pane of the Help window, you might see an additional, even more refined listing of help topics related to troubleshooting a particular object. Select the help topic by clicking it.

In most cases, your request for help is answered by a list of steps to perform a particular action using one of Publisher's features. For instance, Figure 7.3 shows the help provided regarding the editing of clip art with Microsoft Draw.

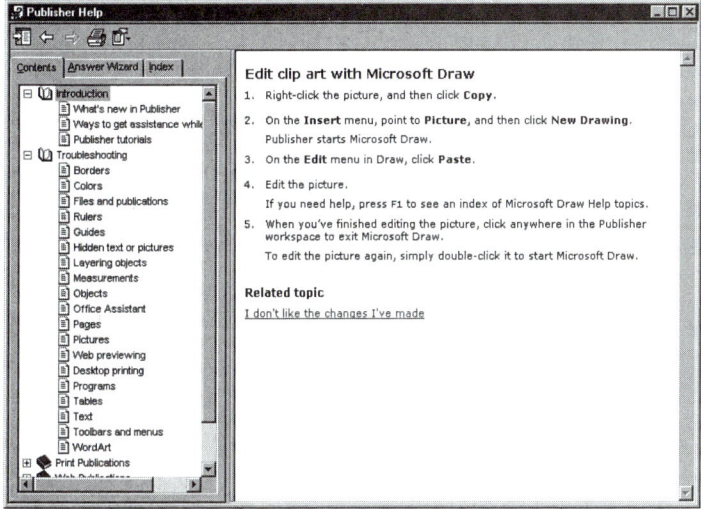

FIGURE 7.3 Specific help is provided in the right pane of the Help window, usually in the form of a series of steps.

USING THE ANSWER WIZARD

Another way to get help in the Help window is to use the Answer Wizard. The Answer Wizard works exactly the same way as the Office Assistant does; you ask the Wizard questions, and it supplies you with a list of topics that relate to your question. You click one of the choices provided to view help in the Help window.

To use the Answer Wizard, follow these steps:

1. Click the Answer Wizard tab in the Help window. Type your question in the What Would You Like to Do? box.

2. Click the Search button. A list of topics appears in the Select Topic to Display box.

3. Double-click a topic, and help related to the topic appears in the right pane of the Help window (Figure 7.4).

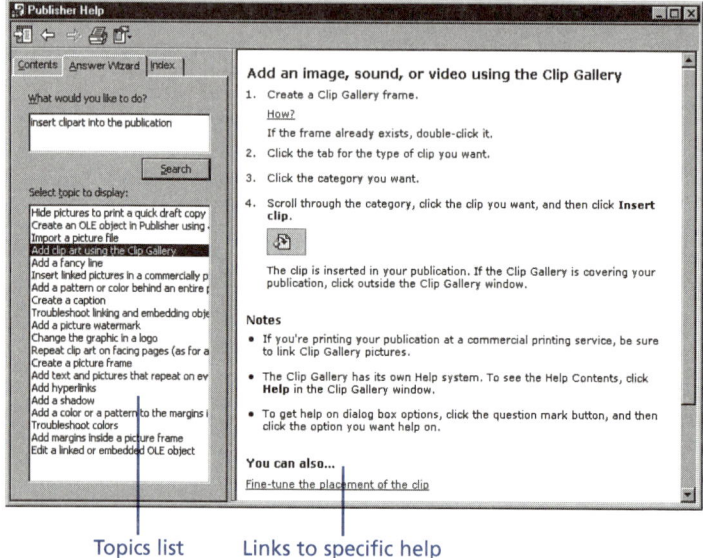

FIGURE 7.4 The Answer Wizard provides you with a list of topics that you can choose from to get help with a particular feature.

USING THE INDEX

Another way to get help from the Publisher Help window is to do key word searches in the Help system's Index. To access the Index system, click the Index tab in the Help window. The Index enables you to type in a keyword or keywords, select from a list of existing keywords, or select from a list of topics that appear based on your keywords or the keywords you choose from the keyword list.

To use the Index, follow these steps:

1. Type a keyword in the Type Keywords box. Notice that as you type, Publisher attempts to complete the keyword for you.

2. After typing in the keyword, click the Search button or select a more appropriate keyword from the keyword list, and then click Search.

3. A list of help topics will appear in the Choose a Topic box. Select the appropriate topic in the list, and the associated help appears in the right pane of the Help window.

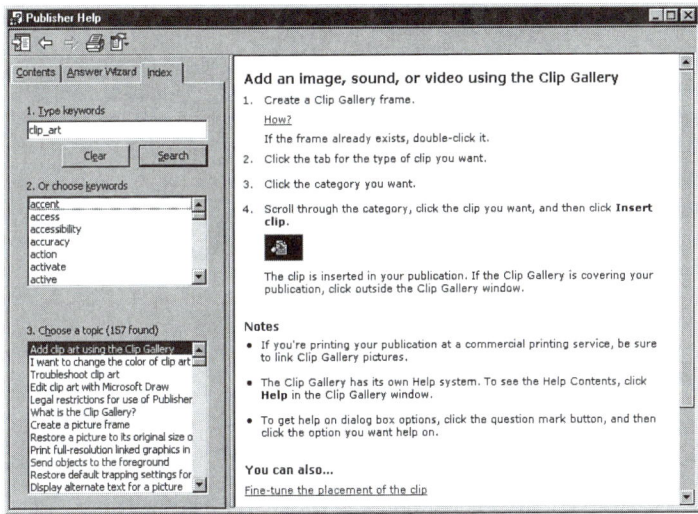

FIGURE 7.5 The Index tab enables you to search for help using keywords.

As with the other forms of help (Contents and Answer Wizard), you can use the hyperlinks provided on the Help screens to move to specific areas of the help system. If you want to begin another keyword search, you can clear the current topics in the Choose a Topic box by clicking the Clear button.

When you complete your search for help using the Index or find help using the Contents or Answer Wizard tab, you can do any of the following:

- Print the topic by clicking the Print button.
- Click the Back button to view the previous Help window.
- Hide the Help tabs by clicking the Hide button.
- Click the Close (×) button to close Help.

What's This?

In most of its dialog boxes, Publisher includes a handy tool called a What's This (?) button. This is the question mark button that appears at the right end of the title bar next to the Close (×) button. You can click the What's This button and then click any item in the dialog box to see an explanation or definition of the selected element. Click again to hide the Help box.

Another way to use this feature is to open the Help menu and choose the What's This command. When the mouse pointer changes to a pointer with a question mark, click anything in the Publisher window that you have a question about, and a Help box explaining the item appears.

Getting Help Online

If you want to get more help than the Publisher help system can provide, you can look for additional help online at Microsoft's Office Web site. To get help online, follow these steps:

1. Make sure that you are connected to your Internet service provider.

2. Select the Help menu, and then select Microsoft Publisher Web Site. Your Web browser opens, and you are taken to the Microsoft Office Update site.

3. Click the appropriate link to get the help that you need.

4. When you finish working with the online help, close your Web browser to return to the Publisher application window.

In this lesson, you learned to use the Office Assistant, to search for help on specific topics, and to use other Help features. In the next lesson, you learn to work with publication frames.

Lesson 8

Working with Publication Frames

In this lesson, you learn the basics of working with frames and inserting, copying, deleting, and manipulating frames on your publication pages.

Inserting a Frame

When you place an item, such as text, a picture, or another object, on a publication page, you are actually inserting a *frame* that contains the particular item. Being able to insert, delete, or move frames and manage their border and color attributes provides you with the capability to give your publication pages a customized look and control the overall layout of individual pages.

 Frame The bordered space that holds your text, picture, or other item.

To place a frame on your page follow these steps:

1. Click a frame tool on the Publisher toolbar (such as the Text Frame tool).

2. Place the mouse pointer on the page where you want to place the new frame. The mouse pointer becomes crosshairs (see Figure 8.1).

3. Click and drag to create the frame as shown in Figure 8.1 (you determine the height and width of the frame).

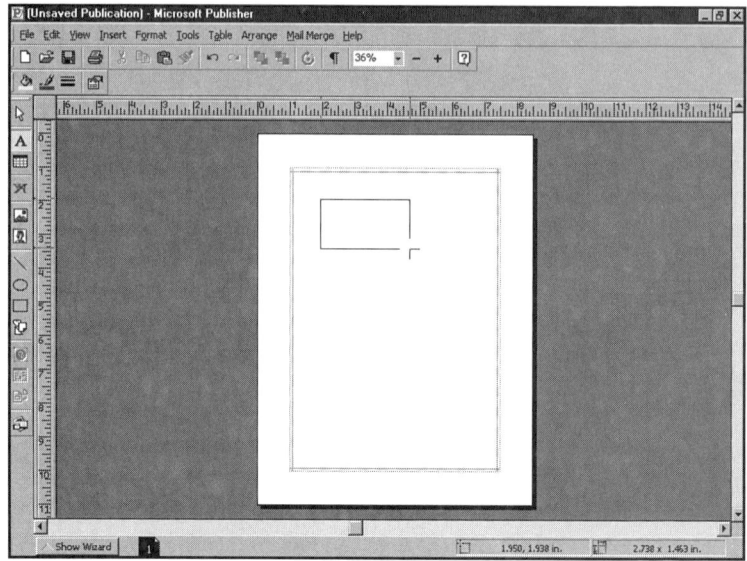

FIGURE 8.1 Click and drag to place the new frame on your publication page.

The new frame appears on your page. Your next action depends on the type of frame you created. If you used the Text Frame tool to create the frame, you now type the text you want to place in the frame (For more about working with text, see Lesson 10, "Changing How Text Looks.") If you used the Picture Frame tool, the Insert Clip Art dialog box appears, enabling you to insert your choice of pictures. (For more about working with pictures and clip art in Publisher, see Lesson 11, "Working with Graphics.")

When you have the frame on the page, a number of options are available to you. You can size the frame, move the frame, delete the frame, or group the frame with other frames on the page. These frame manipulations are covered in the balance of this lesson.

Removing a frame from a publication page is very straightforward. Select the frame that you want to delete, and then press the **Delete** key on the keyboard. This removes the frame from the publication.

Undoing the Placement of a New Frame If you place a frame on a page and want to quickly remove it, click the Undo button on the standard toolbar.

SIZING A FRAME

You can change the width, height, or both of a frame that is located either on a publication page or in the background of your publication (see Lesson 16, "Formatting Publication Pages"). Changing the size of a frame is accomplished using the sizing handles that appear on the selected frame.

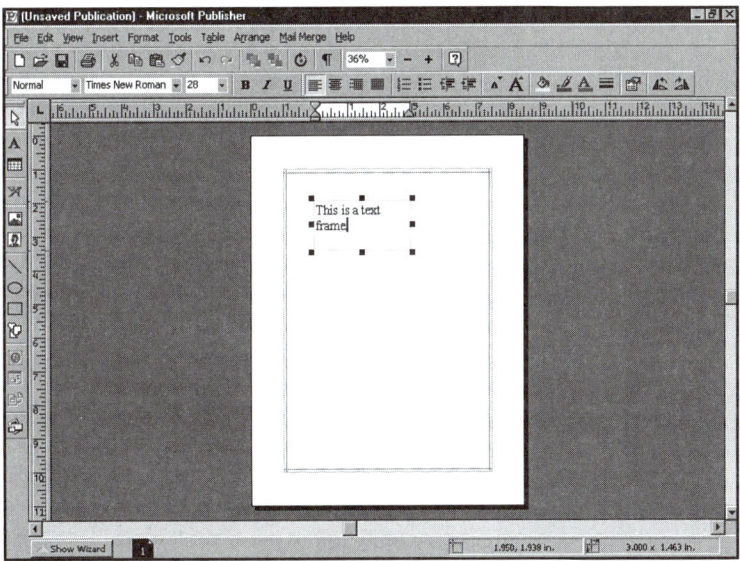

FIGURE 8.2 The sizing handles provide you with a way to change the height or width of a selected frame.

Select the frame by clicking the object that is contained in the frame. Sizing handles appear on the border of the frame (see Figure 8.2). To change the frame size, select one of the options discussed in the following bulleted list:

- **Change the Width** To change the width of the frame, place the mouse pointer on one of the sizing handles on either of the vertical borders (left or right) of the frame. The mouse pointer changes to a Resize pointer. Drag to change the width of the frame.

- **Change the Height** To change the height of the frame, place the mouse pointer on one of the sizing handles on either the top or bottom horizontal border of the frame. Drag to change the width of the frame using the Resize pointer.

- **Change the Width and Height** To change the width and height of the frame simultaneously and maintain the current width and height ratio, place the mouse pointer on any of the diagonal sizing handles (handles positioned where the vertical and horizontal border meet in a corner) and drag to change the overall size of the frame (see Figure 8.3).

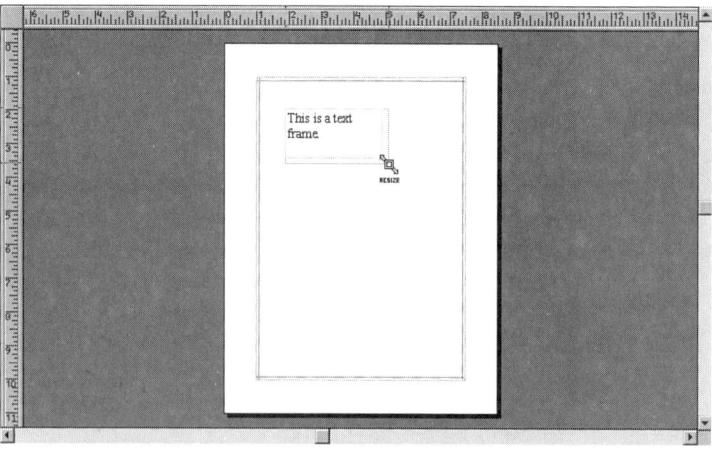

FIGURE 8.3 Drag any of the sizing handles to change the size of the frame.

WORKING WITH PUBLICATION FRAMES   73

  Zoom Out When Sizing a Frame   If you zoomed in on a frame to place text or some other object in it, you might want to zoom out to the Whole Page view when you size your frame. Click View, then Zoom, and then Whole Page to switch to the Whole Page view. Now you can size the frame in relation to other frames on the page.

If you require more exacting measurements than you can attain with the mouse for the height and width of a particular frame, you can also specify these measurements in the Size and Position dialog box. Follow these steps to specify an exact set of measurements for a frame:

1. Click anywhere on a frame to select it.
2. Select the Format menu, and then select Size and Position. The Size and Position dialog box appears (Figure 8.4).

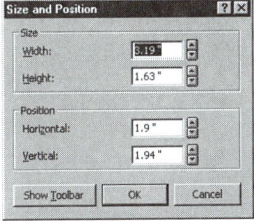

FIGURE 8.4   Use the Size and Position dialog box to control the size of your selected frame.

3. Use the click arrows or type a new width in the Width box.
4. Use the click arrows or enter a new height in the Height box.
5. Click OK to resize the frame and close the dialog box.

Your frame is resized using the entered width and height values.

Control Frame Size with the Measurements Toolbar
You can also control the size and position of a frame using the measurements toolbar. Select View, then select Toolbars, and select measurements from the toolbar list. The second set of measurement boxes (from the left) on the measurements toolbar controls the width and height of the selected frame.

MOVING A FRAME

Publisher also provides you with the capability to move your frames on your publication pages. Any selected frame can be moved using the mouse or the Size and Position dialog box.

Follow these steps to move a frame with your mouse:

1. Click anywhere on a frame to select it.

2. Place the mouse pointer on any of the border edges surrounding the frame (do not place the mouse pointer on the sizing handles). A Move pointer appears.

3. Drag the frame to a new position on the page.

You can also place a frame in a particular position on the page using the Size and Position dialog box; follow these steps:

1. Select Format, and then select Size and Position to open the dialog box.

2. Use the Horizontal Position box to select the horizontal position for the frame, and use the Vertical Position box to set the vertical position.

3. After you enter your position settings, click the OK button.

> **Snapping Frames to Grid and Ruler Guides** Another way that you can position frames with more accuracy is to turn on the Snap to Grid feature. Select Tools, and then Snap to Grids. Frames now snap to the nearest grid line. For more about guides, see Lesson 6, "Working with Existing Publications."

You might find that you want to fine-tune the position of a frame in reference to other frames and objects on a page. This can be done using the Nudge feature:

1. Click anywhere on a frame to select it.

2. Select the Arrange menu, and then select Nudge. The Nudge dialog box appears (Figure 8.5).

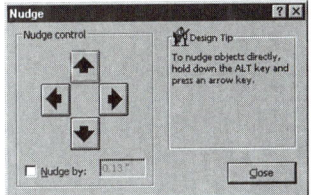

FIGURE 8.5 Use the Nudge dialog box to move a frame slightly in any direction.

3. Click the appropriate arrow button in the Nudge dialog box to move the frame in that direction.

4. To nudge the frame by a specific measurement, click the Nudge by check box and enter a number in the accompanying text box. When you click the arrow buttons on the Nudge dialog box, the frame moves in that direction (by the increment specified).

5. When you finish nudging the object (you can drag the dialog box out of the way to view the frame), click the Close button to close the Nudge dialog box.

Copying a Frame

You can also copy frames and place multiple occurrences of the same frame on a page or copy a frame to another page in your publication. This enables you to easily place repeating design elements on a page or within an entire publication.

To copy a frame, follow these steps:

1. Click a frame to select it.

2. Select the Edit menu, and then select Copy.

3. Select the page from the status bar that you want to place the copy of the frame on, or remain on the current page.

4. Select the Edit menu, and then select Paste.

If you want to move the frame from the current page to another page in the publication, select the Edit menu, select Cut, and then proceed with steps 3 and 4.

> **Quickly Copy and Paste Using the Toolbar** You can also copy, cut, and paste using the toolbar. Click the Copy button or the Cut button, and then select the Paste button, after moving to the appropriate page.

Grouping Frames

After you place frames on a page, you might want to adjust the overall positioning of all the frames in relation to the top or bottom of the page or some other special element on the page (such as a large banner heading). Moving each of the frames individually can be time-consuming and frustrating, especially if you have the frames currently positioned exactly where you like them in relation to each other.

The solution to this problem is to group the frames and then move them together as one unit. This enables you to fine-tune the layout of the page without moving each frame individually.

To group frames, follow these steps:

1. Select the first frame to be in the group by clicking it.
2. Hold down the **Shift** key and select additional frames. Notice that a selection box appears around all the selected frames as shown in Figure 8.6. A Group Objects icon also appears on the selection box.

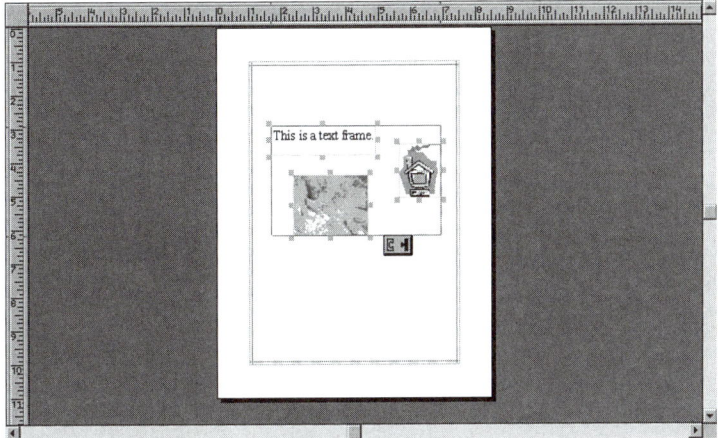

FIGURE 8.6 Group several frames and then move them together to a new position on the page.

You can now move the entire group of frames by clicking the group frame and dragging it to a new position. You can also delete all the selected frames or copy and paste them on another page in your publication.

When you finish manipulating the grouped frames, click anywhere outside the group to make the group frame disappear.

If you want to group the objects on a more permanent basis (keep them together as a group), select the frames to be part of the group, and then select Arrange and Group Objects. Even when you click outside these grouped frames to deselect them, the group remains intact. Click any frame in the group, and all the frames are selected within the group frame. Selecting the Ungroup Objects command on the Arrange menu ungroups the frames in a selected group.

> **Arrange Frame Groups Using the Align Objects Command** You can also align an entire group of objects (or just one object) using the Align Objects command. Select a group of objects. Select the Arrange menu, and then select Align Objects. In the Align Objects dialog box, use the appropriate option buttons to align the frames from left to right or top to bottom.

ARRANGING FRAMES IN LAYERS

You might find occasion to layer several frames on top of each other in a stack. For instance, you might want to place a text frame on top of a picture to produce an eye-catching heading for a publication. You can layer a number of objects and their frames using the layering commands on the Arrange menu.

To layer frames, follow these steps:

1. Drag a frame onto another frame to form a layer. For instance, drag a text frame onto a picture frame.

2. The text in the text frame seems to disappear. With the text frame still selected, click the Arrange menu, and then click Bring to Front. The text frame is placed on top of the picture frame as shown in Figure 8.7.

With some practice, you can layer several frames into complex arrangements on your publication pages. Understanding the layering commands on the Arrange menu can help you work with the frames that you layer. See the following Arrange menu commands:

- **Bring to Front** This moves the currently selected frame to the top of the layered stack of frames.

- **Send to Back** This moves the currently selected frame to the bottom of the stack.

- **Bring Forward** This moves the currently selected frame up one position in the stack. For instance, if the frame is the second frame in the stack of the layers, this moves the frame to the first position, or top, of the stack.

- **Send Backward**   This moves the currently selected frame down one position in the stack. For instance, a frame in the second layer of the stack is moved to the third layer of the stack.

FIGURE 8.7   Layer frames on your pages to make your publications interesting.

 Make Your Frame Stack One Group   When you layer several frames and position and stack them appropriately, make the frames a permanent group (select Arrange, and then Group Objects). This precludes you from inadvertently disturbing the stack when you are working on the other elements on the publication page.

In this lesson, you learned to insert, size, move, and work with frames on your publication pages. In the next lesson, you learn to enhance your frames with borders and colors.

Lesson 9

Enhancing Frames with Borders and Colors

In this lesson, you learn to create borders and add colors and shading to the frames on your publication pages.

Adding Borders to Frames

When you place a frame on a publication page, the frame is transparent and does not have a border. You can really only discern the size and shape of the frame when you select the object (for instance, the text or picture) that the frame contains. You can add border lines to your frame and also place shading and background colors on any frame you create.

Adding borders to your frames is very straightforward. You have control over the line thickness, style, and color.

To add borders to a frame, follow these steps:

1. Click on the frame you want to place the border around.
2. Click the Format menu, and then click Line/Border style.
3. Click a Line Style on the cascading menu that appears. A border appears around your frame.

You don't have to always place a box around a selected frame. You can select certain borders of the frame (such as the left vertical border or the bottom horizontal border) and add a border line to this area of the frame. This is done in the Border Style dialog box.

Follow these steps to add selected borders to a frame:

1. Select Format, then Line/Border Style, and then More Styles. The Border Style dialog box appears.

2. Use the Select a Side box in the dialog box to select certain sides of the frame for formatting (see Figure 9.1). For instance, click the top of the frame in the Select a Side box, and the border indicator triangles for the rest of the border disappear.

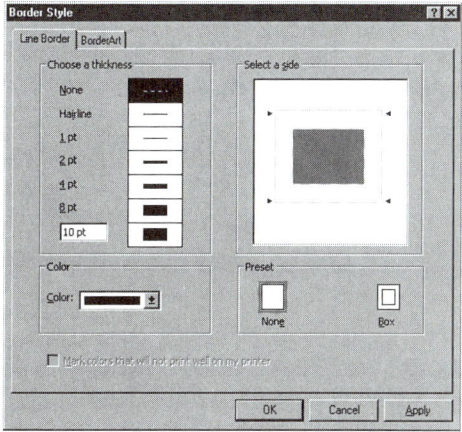

FIGURE 9.1 Select the sides you want to place a border on in the Border Style dialog box.

3. Hold down the **Shift** key, and click on any of the other frame sides you also want to place a border on.

4. Click a border thickness on the left side of the dialog box.

5. Click OK to add the border to your frame.

CHANGING BORDER ATTRIBUTES

When you have a border around a particular frame, it is very easy to change that border's attributes, such as color or line style. In fact, when you select a frame, buttons related to border attributes appear on the Publisher formatting toolbar. Table 9.1 provides a thumbnail sketch of each frame formatting button.

TABLE 9.1   THE FORMATTING TOOLBAR BUTTONS FOR EDITING BORDER ATTRIBUTES

TOOLBAR BUTTON	FRAME ATTRIBUTE
	Fill Color
	Border Color
	Line/Border Style

CHANGING BORDER COLOR

When you assign a border to a frame, you can change the color of that border. Follow these steps:

1. Select the frame.

2. Click the Border Color button on the formatting toolbar (see Figure 9.2).

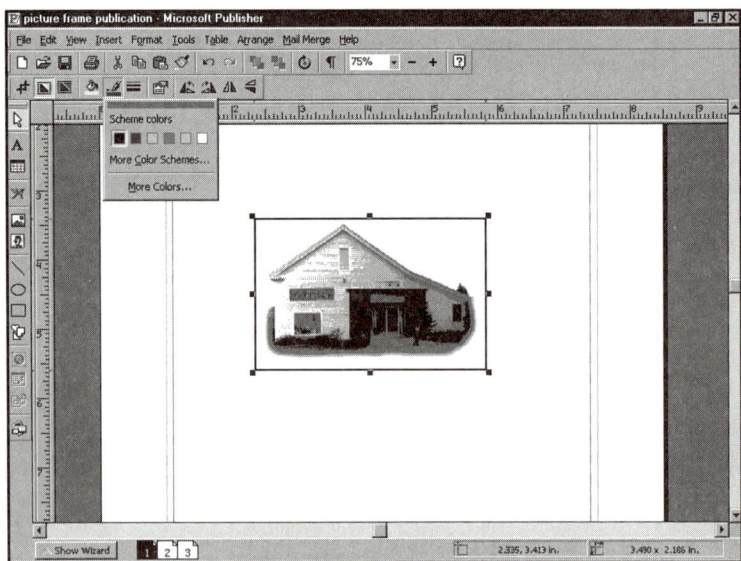

FIGURE 9.2   Use the Border Color button to assign a new color to your frame's border.

3. Click a color on the Border Color box.

The border around your frame is assigned the selected color. If you want to choose from more colors than provided by the Color drop-down box, click the More Colors button on the drop-down list. The Colors dialog box appears. Click in the Color Spectrum box to select a new color for your frame borders. Then click OK to return to the Border Style dialog box.

USING FILL COLORS

You can also fill a particular frame with a selected color. This is a great way to add interest to your publications, particularly those you print in color. When you add color to the frame, be advised that it can obscure the object that resides inside the frame. When you color a text frame, you might have to change the text color so it can still be read after a fill color is selected. For instance, a blue fill color might require a yellow text color for the text to be readable. For more about working with text in frames, see Lesson 10, "Changing How Text Looks."

In the case of picture frames, the fill color fills only the interior part of the frame that is not occupied by the picture itself. The color, in effect, becomes a background for the picture.

To add a fill color to a frame, follow these steps:

1. Select the frame that you want to add a fill color to.
2. Click the Fill Color button on the Formatting toolbar.
3. Select a fill color from the color box that appears. The fill color appears in the selected frame.

If you want to choose from more colors than provided by the Fill Color button, click the More Colors selection on the color box. The Colors dialog box appears. Click in the Color Spectrum box to select a new color for your frame borders. Click OK to place that color in the frame as the fill color.

Using Fill Effects

When you work with fill colors in your frames, you can actually manipulate these colors with tints, patterns, and gradients that really add visual interest to your frames. You work with the various fill effects in the Fill Effects dialog box.

To add effects to a frame's fill color, follow these steps:

1. Select the frame that you want to add fill effects to.

2. Click the Fill Color button on the formatting toolbar.

3. Click Fill Effects on the drop-down box that appears. The Fill Effects dialog box opens (see Figure 9.3).

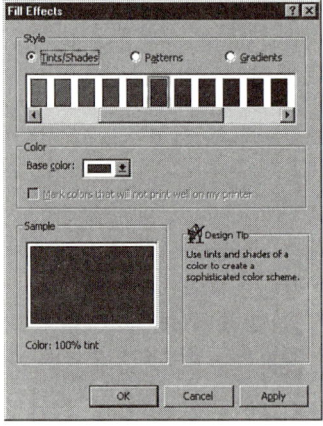

FIGURE 9.3   You can add effects to a frame's fill color in the Fill Effects dialog box.

4. To select a particular fill effect, select one of the option buttons at the top of the dialog box:

- **Tints/Shades**   Select this option button if you want to apply a special tint/shade for the current fill color. You can then select a color that ranges from white through all the tints and shades up to black.

- **Patterns** Select this option button to select from a number of different patterns (including lines, crosshatching, and so on) based on the current fill color.

- **Gradients** Select this option button to apply a particular gradient (swirling and patterned gradations of the current fill color) from a range of supplied gradient effects.

5. After you select a fill effect category, select an effect in the scroll box located directly beneath the effects option buttons. The effect that you select is previewed in the Sample box.

6. When you have selected an effect you wish to apply to your frame's fill, click OK.

The chosen effect formats the fill area of your frame.

Applying Shading

You can also place a shadow on your Frame border. This shading on the bottom and right sides of the frame adds depth to the frame. Shadows particularly enhance text frames and picture frames where you place a border around a particular frame. The shadow "lifts" the frame away from the publication page, giving it a three-dimensional feel.

To add shading to a frame, follow these steps:

1. Click the frame you want to place the shadow on.

2. Click the Format menu, and then click Shadow.

Shading appears on the right and bottom of the frame as shown in Figure 9.4.

Shadows work best when a border is placed around a particular frame. The shadow adds depth and definition to an otherwise flat border and frame.

In this lesson, you learned to place borders around frames and change the color and other attributes of those borders. You also worked with frame fill colors and placed a shadow around a frame. In the next lesson you learn how to change the way text looks in text frames and how to add special text objects such as mastheads to publications.

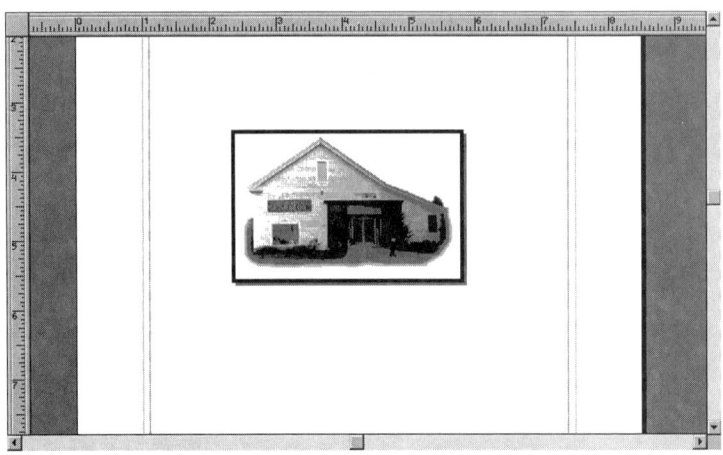

FIGURE 9.4   Shading gives your frames depth.

Lesson 10

Changing How Text Looks

In this lesson, you learn how to add text to your publications and change formatting options related to text, such as font selection, text alignment, and text color. You also learn to add text banners to your pages and connect text frames together.

Adding Text to Your Publications

Text is added to your publication in text frames using the Text Frame tool on the Publisher toolbar. Publisher provides you with complete control over the look and formatting of text frames, including such elements as the font's style, size, attributes (such as bold and italics), and color. Any or all of these font parameters can be edited on a particular text frame.

To add a text frame to a publication page, follow these steps:

1. Click the Text Frame tool on the Publisher toolbar.
2. Position the mouse pointer on the page and drag to create the Text Frame.
3. The insertion point appears in the text frame. Type the text you want to place in the frame.

 Zoom in to Concentrate on Your Text If you are in the Whole Page view when you place your text frames on your page, you might want to zoom in on the text before you type the text or attempt to edit it. Click the Zoom drop-down box on the standard toolbar, and select a zoom percentage that zooms you in on your page.

INSERTING TEXT

You can also easily add text to a text frame. To add text to a frame that already contains text, follow these steps:

1. Place the mouse pointer on the text in the text frame. The pointer becomes an I-Beam.

2. Click the I-Beam on the text where you want to place the insertion point.

3. Type the text you want to insert in the text box (see Figure 10.1).

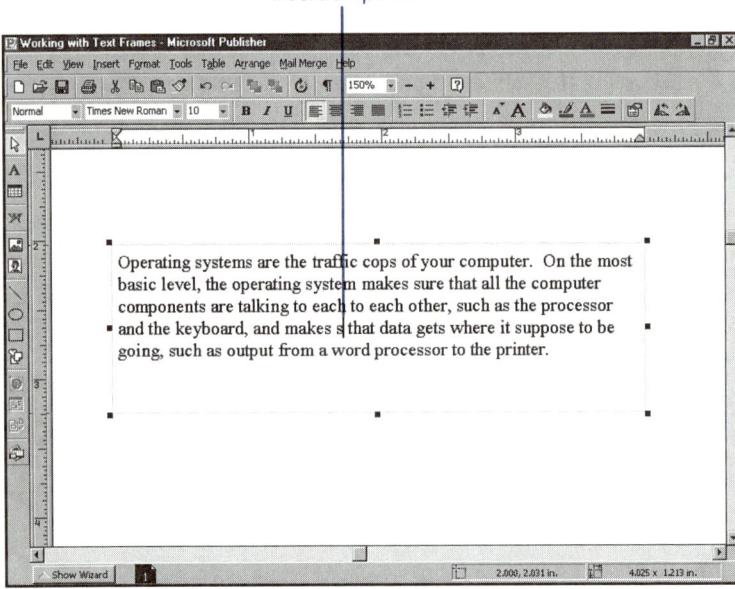

FIGURE 10.1 Place the insertion point in a text box, and then use the keyboard to add text.

SELECTING AND DELETING TEXT

You can also easily delete text from a text frame. It's just a matter of selecting the text that you want to delete and then pressing the **Delete** key on the keyboard. The mouse provides the easiest method for selecting text in the text frame. Table 10.1 provides a list of different ways to use the mouse to select text.

TABLE 10.1 MOUSE CLICKING TO SELECT TEXT

TEXT SELECTION	MOUSE ACTION
Selects the word	Double-click on a word
Selects the word	Double-click on a word
Selects text block	Click and drag
	Or click at beginning of text and hold down **Shift** key and click at the end of text block
Selects all the text	Triple-click in the text frame

These selection techniques are also useful when you want to change the format (or color) of text in a text frame (as discussed in subsequent sections of this lesson).

Copy, Cut, and Paste Text   You can also copy or cut selected text in a text box and then paste it into another text frame or in another position in the current text box. Use the Copy button, the Cut button, and the Paste button respectively on the standard toolbar.

WORKING WITH FONTS

The text that you type in a new frame is created in the default Publisher font, which is Times New Roman, 10 point. Each font available has a particular style or typeface.

A variety of font types exist, such as Arial, Courier, Times New Roman, CG Times, Bookman Old Style, and so on; the fonts you have to choose from depend on the fonts that are installed on your computer. Windows 98 and 95 provide most of the font families that applications such as

Publisher use. Publisher also installs several font types when you install it from the Microsoft Office or the Publisher 2000 CD-ROM (see the Introduction for some installation considerations for Publisher).

The size of the font is measured in points. A point is 1/72 of an inch, with the standard point size for business letters and other documents being 12 point—the higher the number of points (such as 18), the larger the font size.

You can change the font for existing text in a text frame by following these steps:

1. Select the text in the text frame (using the mouse) that you want to change to a different font type.

2. Click the font drop-down box on the formatting toolbar, and select a new font from the list (see Figure 10.2).

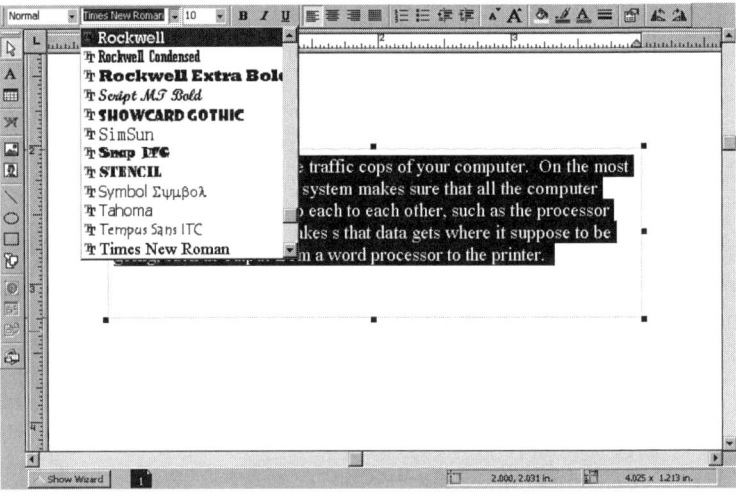

FIGURE 10.2 Choose a new font for selected text using the font drop-down box.

If you finish working with a particular text frame, you can click outside the frame to deselect the frame itself.

If you create a new text frame, you can change the font that you use for the text that you place in the frame before you actually type any text.

This, in effect, changes the default font for that particular text frame. To change to a different font, click the font drop-down box and select a new font.

You can change the font size of text in a text frame in much the same way. Select the text in the text frame that you want to assign a new font size to. Click the font size drop-down box on the formatting toolbar, and select a new font size.

CHANGING FONT ATTRIBUTES

You also have control over a number of other font attributes associated with the text in your text frames. You can quickly change the style of the font to bold, italic, or underline. These font styles are readily available on the formatting toolbar, as shown in Table 10.2. A number of other font attributes can also be selected in the Font dialog box.

TABLE 10.2   FORMATTING TOOLBAR BUTTONS TO CHANGE TEXT STYLES

FORMATTING TOOLBAR BUTTON	STYLE
B	Bold
I	Italic
U	Underline

To change the font attributes for text in a text frame, follow these steps:

1. Select the text in the text frame that you want to change the font attributes for (for example, changing the text style to bold).
2. Click the appropriate button on the formatting toolbar.

As already mentioned, you can also change font attributes for selected text in a frame using the Font dialog box. Select the Format menu, and then select Font.

The Font dialog box enables you to change the font, the font style, the font size, and a number of other font attributes (such as superscript, subscript, small caps, and so on) as shown in Figure 10.3.

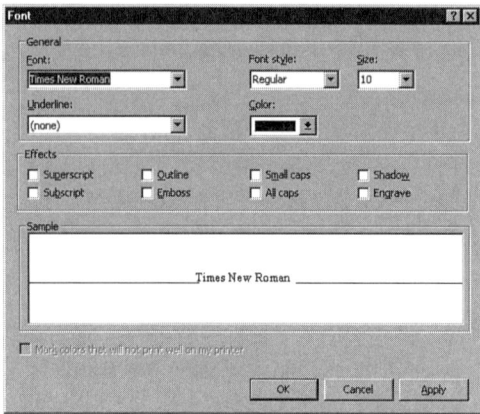

FIGURE 10.3 The Font dialog box gives you control over a number of font attributes.

To select any of the font attributes in the Effects area of the dialog box, click the appropriate check box. A preview of the particular effect (including other attribute changes you have made) appears in the Sample box. When you complete your font changes, click the OK button to return to your publication.

CHANGING FONT COLORS

You can also change the color of the text in a text frame. Changing font color enables you to emphasize certain text and can add interest to your publication pages.

To change the color of text in a text frame, follow these steps:

1. Select the text in the text frame that you want to change to a different font color.

2. Click the Font Color button on the formatting toolbar.

3. Select a new color from the Color box that appears (if you don't see a color you like, continue with Step 4).

4. If you want to select from additional colors, click the More Colors selection in the Color box. The Color dialog box appears (see Figure 10.4).

Changing How Text Looks

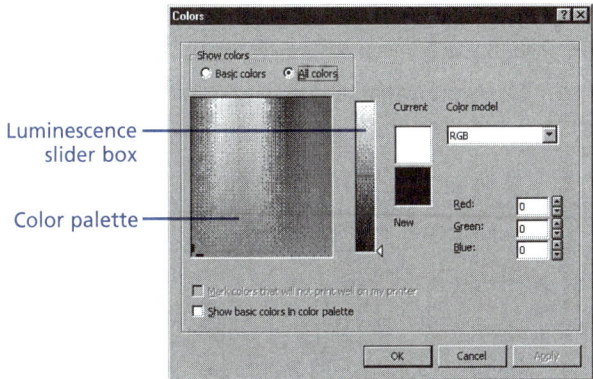

Luminescence slider box

Color palette

Figure 10.4 The Color dialog box offers you all the colors of the spectrum for your text.

5. Click anywhere in the color palette to select a new color range for the text.

6. In the color luminescence box to the right of the color palette, drag the slider to select the depth of the color to use for your text.

7. After you select your new color, click OK to return to your publication.

 Luminescence In Publisher, luminescence is the tone, or degrees of light and dark, in a particular color. The luminescence slider bar enables you to select the tone for the currently selected palette color.

Aligning Text in a Frame

You also have control over the alignment of the text in the frame. You can center the text (in relation to the frame), right-align the text, left-align the text, or use different alignments on different text lines or paragraphs in a text box. For lines or paragraphs to be treated separately, all you have to do is place a line break (press **Enter**) between the line and the next line in the text frame.

The available alignments align the text in respect to the left and right borders of the frame. Table 10.3 gives you a summary of the alignment buttons available on the formatting toolbar.

TABLE 10.3 FORMATTING TOOLBAR BUTTONS TO ALIGNMENT TEXT IN FRAMES

FORMATTING TOOLBAR BUTTON	ALIGNMENT
	Left
	Center
	Right
	Justify (straight margins on the left and right).

To align text in a frame, follow these steps:

1. Place the insertion point in the paragraph (any line or lines of text followed by a line break) that you want to align.

2. Click the appropriate button on the formatting toolbar (as detailed in Table 10.3).

Your text is aligned according to the button you selected on the formatting toolbar.

Align Multiple Paragraphs If you have a text box with more than one paragraph of information in it, you can select all or some of the paragraphs and then click an alignment button to change the alignment of several paragraphs at once.

ADDING TEXT MASTHEADS

Publisher does not limit you to text frames for the text items that you place on your publication pages. You can also add special text banners using the Publisher Design Gallery. These text banners, called *mastheads* in Publisher, are predesigned elements that offer color and other design items that really help dress up text entries in your publications. Mastheads work great as the heading text for your publication pages.

To add a masthead from the Design Gallery, follow these steps:

1. Click the Design Gallery button on the Publisher toolbar. The Design Gallery appears.

2. Make sure that Mastheads is selected in the Categories pane of the Design Gallery (see Figure 10.5).

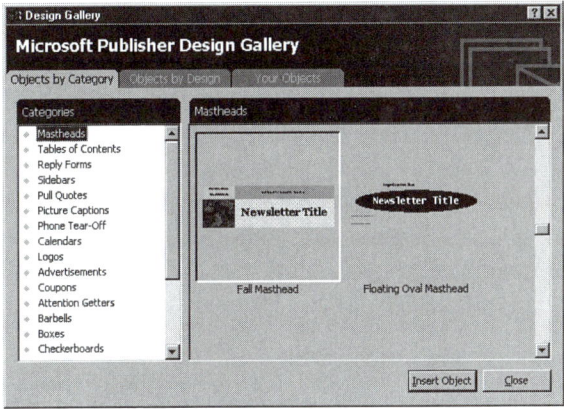

FIGURE 10.5 You can add mastheads to your publications as page headings.

3. Scroll through the mastheads available in the Mastheads pane of the Design Gallery, and select a masthead for your page.

4. Click the Insert Object button to add the selected masthead to your page.

When the masthead is placed on the page, it contains generic placeholder text. You can edit this text by selecting the text as you would in any text frame and typing your new text to replace it. The text you replace the placeholder text with takes on the placeholder's font and font attributes.

> **Mastheads Are Made of Multiple Frames** Mastheads can be made of several text frames that are formatted differently and grouped together. This means that in any one masthead, you might have to edit several grouped text frames to replace all the placeholder text in the masthead. Just select each individual placeholder text entry, and then type in your text. For more about grouping frames, see Lesson 8, "Working with Publication Frames."

CONNECTING TEXT FRAMES

You might want to place a large amount of text on one publication page, but not in just one text frame (a lot of text stuck in one frame can be overwhelming to the eye and diminish your message). Instead, you can break up the text into more "bite-sized" pieces of information by making the text flow between several text frames as a story. The amount of text in each frame is then determined by the frame size. A story can be placed in several text frames on one page, or the story can flow to text frames that are on different pages of the same publication.

> **Story** Text that flows between several different text frames.

To create a story in multiple text frames, follow these steps:

1. Click the Text Frame tool on the Publisher toolbar, and place a text frame on your publication page. Make the text box big enough to hold all the text in the story.

2. Type all the text to be included in your story in the current text box.

3. Size the text frame (place the mouse on the border of the frame and drag) so that it only shows the text you want to appear in this particular frame (other text is moved to other text frames).

3. When you reduce the size of the text frame, the text you typed exceeds the size of the current text box and an Overflow indicator appears at the bottom of the frame (see Figure 10.6). This means that text not shown in the text frame is being held in an overflow area.

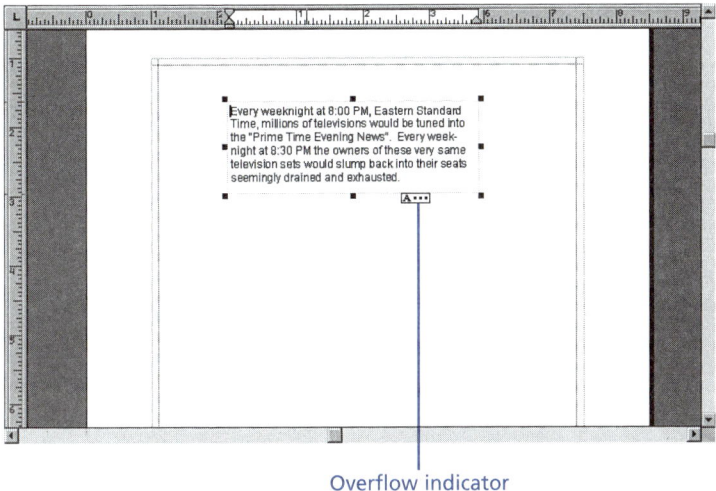

Overflow indicator

FIGURE 10.6   Text that exceeds the size of a text frame is held in an overflow area and an Overflow indicator appears on the text frame.

4. Create the other text frames to hold the overflow portion of the story from the first text frame.
5. Click the text frame that holds the text.
6. Select the Tools menu, and then click Connect Text Frames.
7. A Connect Text Frames button appears on the standard toolbar (see Figure 10.7). Click the Connect Text Frames button.
8. The mouse pointer becomes a pitcher of text as shown in Figure 10.7. Place the pitcher pointer on the first empty text frame that you want to flow the story text to. Click the left mouse button.

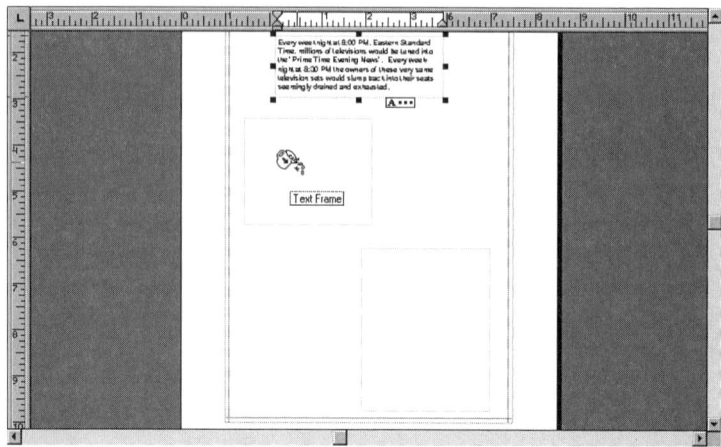

FIGURE 10.7 Select all the text frames that you want to hold the story text.

9. Story text flows into the text frame, and the text frame becomes the selected text frame (a border and sizing handles appear around the text frame). An Overflow indicator might (depending on the amount of text left) also appear on this text frame.

10. Click the Connect Text Frames button on the toolbar again, and use the Pitcher pointer to flow text into the next text frame you have on the page.

When all the text in the original story is placed in the text frames, the last text frame does not have an Overflow indicator on its border. This assures you that all the story text appears in the text frames included in the set. It is best to spell-check large stories. For information on using the spell-checker, see Lesson 17, "Fine-Tuning Publisher Publications."

In this lesson, you learned to change the look of text in a text frame by changing the font's type, size, color, and other attributes such as font style (bold, italic, and so on). You also learned to insert mastheads from the Publisher Design Gallery and create text stories that flow to multiple text frames. In the next lesson, you learn to insert and modify pictures and graphics in your publications.

Lesson 11
Working with Graphics

In this lesson, you learn how to add pictures and clip art to your publications. You also learn how to scale, crop, and change the color of graphic objects, such as pictures and clip art.

Inserting a Picture

Publisher provides you with a lot of flexibility as to the type of objects that you can add to your publication pages. You can add a picture file to a page, or you can choose to add a clip art image from the extensive Clip Gallery that comes with Microsoft Publisher.

Pictures can come in a variety of file types, and pictures can consist of files that you have on disk, items that you copy from the World Wide Web, or pictures that you create using a scanner or a digital camera (for more about inserting pictures from scanners and other sources, see Lesson 12, "Adding Special Objects to Your Publications").

Publisher supports a very wide variety of picture file formats that you can insert in a publication. Table 11.1 lists some of the most common picture file types.

Table 11.1 Publisher-Supported File Formats

File Type	Extension
Windows Bitmap (Windows Paint)	.bmp
CorelDRAW!	.cdr
Encapsulated PostScript (Quark Express)	.eps
Graphics Interchange Format (CompuServe format)	.gif

continues

TABLE 11.1 CONTINUED

FILE TYPE	EXTENSION
Joint Photographic Experts Group (commonly used on the World Wide Web)	.jpeg or .jpg
Kodak Photo CD and Pro Photo CD	.pcd
PC Paintbrush	.pcx
TIFF, Tagged Image File Format (PhotoDraw)	.tif
Windows Metafile (Microsoft Word Clip Art)	.wmf
WordPerfect Graphics	.wpg

To insert a picture on a publication page, follow these steps:

1. Click the Picture Frame tool on the Publisher toolbar.

2. Place the mouse pointer on the page and click and drag to create the Picture frame.

3. With the Picture frame selected, select the Insert menu, and then click Picture. Select From File on the cascading menu.

4. The Insert Picture dialog box appears. Use the Look in drop-down arrow to select the drive that the picture resides on.

5. Double-click the appropriate folder that holds the file.

6. Click the picture file name in the dialog box. A preview of the picture appears (see Figure 11.1).

7. Click the Insert button. The picture is placed in the Picture frame on the publication page.

Open the Insert Picture Dialog Box with a Double-Click You can also insert a picture file into a Picture frame by double-clicking the empty Picture frame. The Insert Picture dialog box appears.

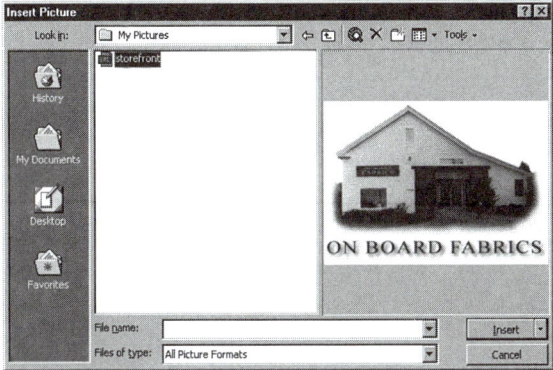

FIGURE 11.1 Use the Insert Picture dialog box to locate the picture file you want to place in your Picture frame.

Using Clip Art

An alternative to placing picture files in your publications is to use clip art. The Clip Gallery is a library of ready-made images that are arranged in categories (such as Animals or Business) to make it easy for you to find the right type of graphic image.

The Publisher Clip Gallery is quite extensive and has clip art that serves in nearly every possibility. To insert clip art on a publication page, follow these steps:

1. Click the Clip Gallery tool on the Publisher toolbar.

2. Place the mouse pointer on the page and drag to create the clip art frame. The Clip Gallery window appears.

3. The Clip Gallery has three tabs: Pictures, Sounds, and Motion Clips. Make sure the Pictures tab is selected. The Categories of clip art available appear in the Clip Gallery window (see Figure 11.2).

4. Click a category (such as Animals) to see the clip art available in that particular category.

Maximize the Gallery Window It is easier to view and navigate the categories in the Clip Gallery if you maximize the Gallery window by clicking the Maximize button in the upper-right corner.

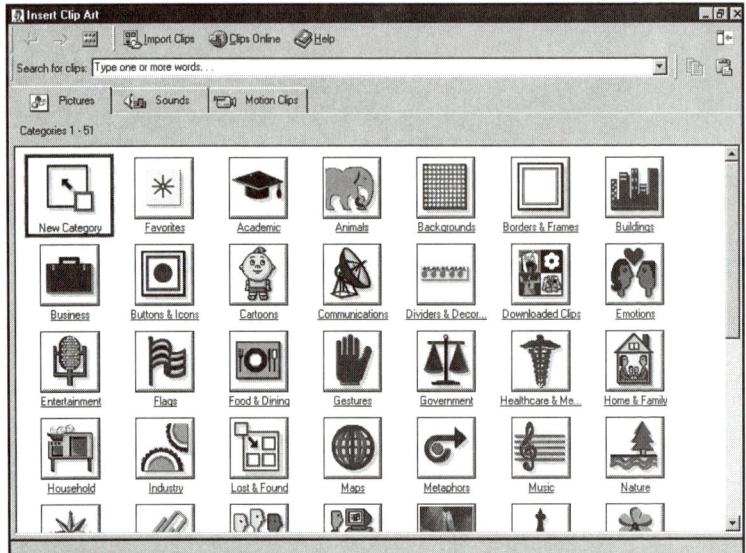

FIGURE **11.2** The Clip Gallery provides you with clip art categories so you can choose images by theme.

5. To insert a particular clip art image, click the image. A toolbar appears next to the image (see Figure 11.3).

6. Click the Insert clip button, and the image is inserted in your publication.

7. When you finish working with the Clip Gallery, click its Close (×) button to return to your publication.

WORKING WITH GRAPHICS 103

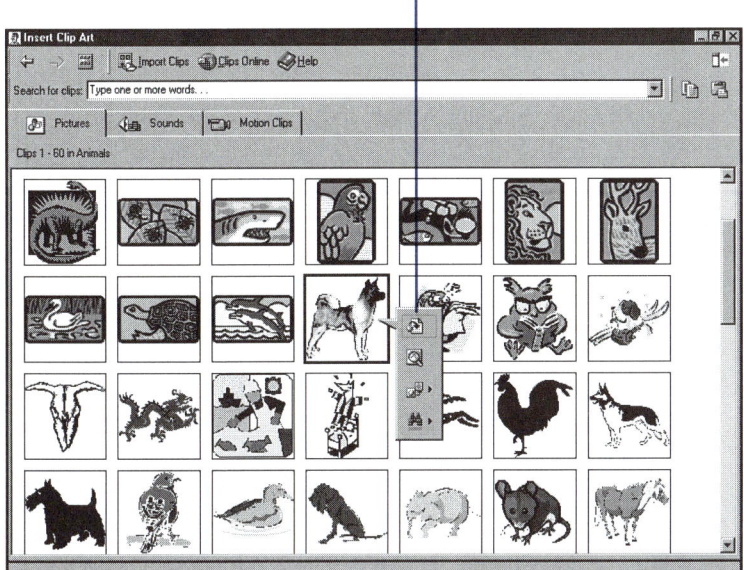

FIGURE 11.3 Click the Insert clip button on the image toolbar to insert the clip art.

 Search for Clips You can also do a search for clip art by keywords. The clips that satisfy the search parameters appear in the Gallery window. Click in the Search for clips box at the top of the Gallery window. Type in your search word and then press **Enter**.

Clip art reacts to changes in the frame size exactly the same way that pictures do. Increase the size of the frame, and the clip art inside the frame is enlarged.

The Clip Gallery also provides you with sound and animated video clips that you can place in publications that are viewed online, such as Web pages. See Lesson 12 for more information.

SCALING PICTURES

When you have a picture in a Picture frame or a clip art image in a clip art frame, you can easily scale the image using the mouse. Select the frame, and then use the sizing handles that appear to size the image.

To increase or decrease the size of an image, drag the image frame (see Figure 11.4).

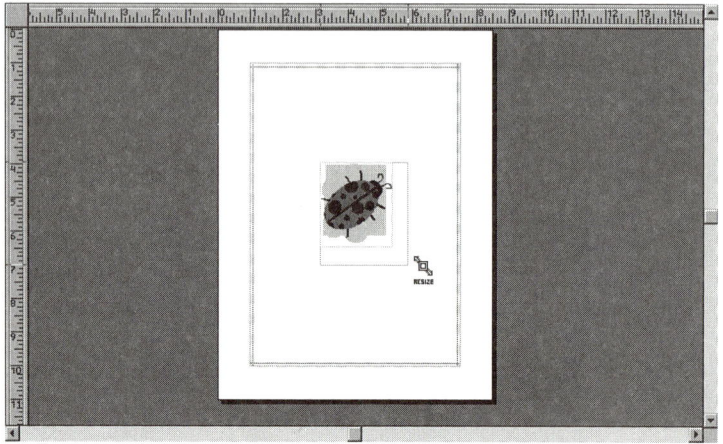

FIGURE 11.4 Drag a sizing handle to scale the image in a particular frame.

 Don't Distort Your Pictures When scaling your pictures, you typically want to maintain the height/width ratio of the picture. Images appear elongated or squashed if you only change their scale in one direction (such as dragging only one of the vertical borders to change the width).

You can also scale an image based on a percentage of the original size of the image. When you insert the image, it fills the frame that you created for it. The frame constitutes a particular width and height for the image. The final size of the inserted image can be less than that of the original image you used (the size it was originally created at and then saved as a file)—less than 100%—or it can be greater than the original image size if you used a particularly large frame—greater than 100%.

Your overall image size is based on two measurements: height and width. You can adjust both of these by a percentage if you want, using the Scale Picture dialog box.

Follow these steps to scale the picture using percentages for width and height:

1. Select the frame that holds the image you want to scale.
2. Select Format, and then select Scale Picture. The Scale Picture dialog box appears.
3. Click in the Scale height box and enter a percentage for the height of the image.
4. Click in the Scale width box and enter a percentage for the width of the image.
5. Click OK to close the dialog box.

The image is scaled according to the percentages you enter. This also, in effect, scales the frame that the picture lives in.

CROPPING PICTURES

You might run across a situation where you want to trim the edges off of a particular image. For instance, you have a clip art image that contains a picture surrounded by a border, and you want to crop the border to keep the only picture in the frame. Another instance is if you have an image that contains several items, such as a picture of several people, and you want to crop the image so only one person appears in the frame.

You can easily crop an image using the Publisher cropping tool. To crop a picture or clip art image, follow these steps:

1. Select the picture or clip art frame that holds the image you want to crop.
2. Select Format, and then select Crop Picture.

3. The mouse pointer becomes the Cropping tool (see Figure 11.5).

4. Place the Cropping tool on any of the sizing handles on the picture's frame and drag to crop the picture as shown in Figure 11.5.

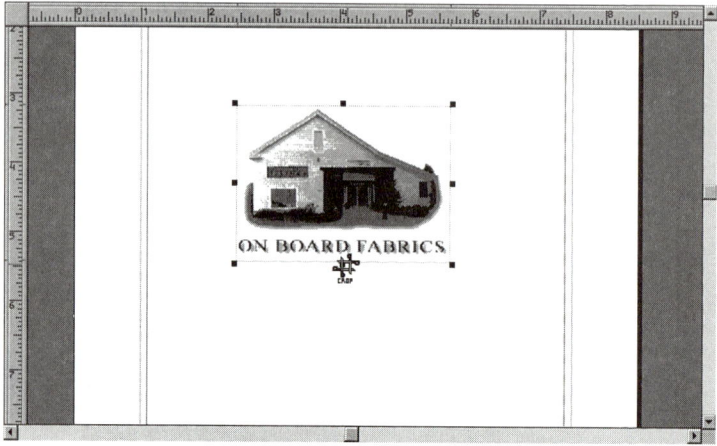

FIGURE 11.5   Use the Cropping tool to crop an image in a picture or clip art frame.

5. When you finish cropping the picture, click anywhere on the page to deselect the image frame.

> **Crop in More than One Direction**   When you invoke the Cropping tool, you can crop the image in more than one direction (such as crop it on the right and then crop it on the top) in the same cropping session. The Cropping tool only disappears when you clip outside the Picture frame and deselect it.

CHANGING PICTURE COLORS

You can also change the color of an image in a picture or clip art frame. Changing the color of a picture means that you change all the current colors to one selected color. Changing picture colors works well with images that are only one or two colors. Images that are multicolored do not look very good when changed to just one color. Coloring images is not without its place, however. You can get some very nice visual effects coloring certain images in your publications.

To change the color of an image, follow these steps:

1. Select the frame that holds the picture or clip art you want to recolor.

2. Click Format, and then click Recolor Picture. The Recolor Picture dialog box appears (see Figure 11.6).

3. Click the Color drop-down box to select a new color (if you find a color you want to use on the color box that appears, you can skip step 4).

FIGURE 11.6 The Recolor Picture dialog box enables you to choose one color for the selected picture.

4. If you want to select from additional colors, click the More Colors selection in the color box. The Color dialog box appears. Click any color on the palette, and then click OK to return to the Recolor Picture dialog box.

5. Click a color on the Color drop-down box and then click OK to close the Recolor Picture dialog box.

The new color is applied to your picture. Because you are creating a monochrome picture (a picture of one color) you might have to experiment a little bit to find a color that gives you a satisfactory image.

In this lesson, you learned to insert pictures and clip art into your publications. You also learned how to scale, crop, and color an image. In the next lesson, you learn how to add special objects, such as banners and other design elements, to your publications.

Lesson 12

Adding Special Objects to Your Publications

In this lesson, you learn how to add special objects using the Design Gallery; you import objects from other applications; you also insert video, audio, and pictures from scanners and cameras.

Using the Design Gallery

Publisher makes it easy for you to add a variety of special objects to your publications. The Design Gallery provides a set of wizards that create *smart objects* (such as logos, mastheads, calendars, and other special objects). These objects are called smart objects because they can be edited at any time using the appropriate wizard, making them very different from the typical static object, such as a picture or clip art, that you add to your pages.

To insert a Design Gallery object onto a publication page, follow these steps:

1. Click the Design Gallery Object tool on the Publisher toolbar.
2. In the Design Gallery window, click an Object category in the Categories pane (such as Calendars).
3. In the Object pane, click the object you want to insert on the page (see Figure 12.1).
4. Click the Insert Object button to insert the object on the page.

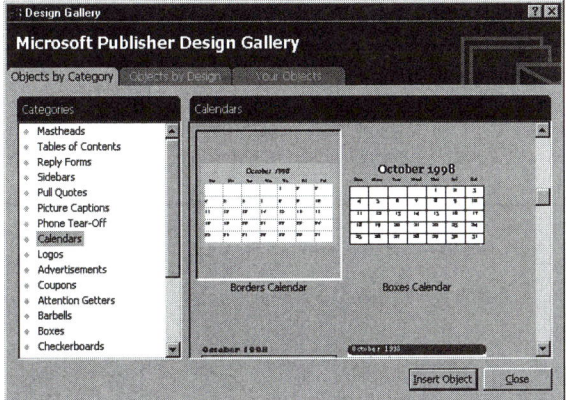

FIGURE 12.1 Insert any number of special objects using the Design Gallery.

Dealing with Oversized Objects   Some of the smart objects you create using the Design Gallery are inserted on your page at a size that cannot be accommodated by a typical page size. You have to size the new object to fit within the boundaries of your page.

Editing Design Gallery Objects

When you add a smart object to your publication using the Design Gallery, Publisher makes it very easy for you to change the formatting of any of these special items. A wizard is associated with each smart object and can be consulted to make changes to a particular object. For instance, you can create a calendar, such as the one that is shown in Figure 12.2, and change the look of it.

To edit a Design Gallery Object, follow these steps:

1. Click the smart object to select it.
2. Click the Wizard button that appears on the smart object's border (it contains a magic wand). The wizard associated with the object opens (such as the Calendar Creation Wizard shown in Figure 12.2).

3. Click a design element category in the top pane of the wizard (in this case, Design is selected in the Calendar Creation Wizard).

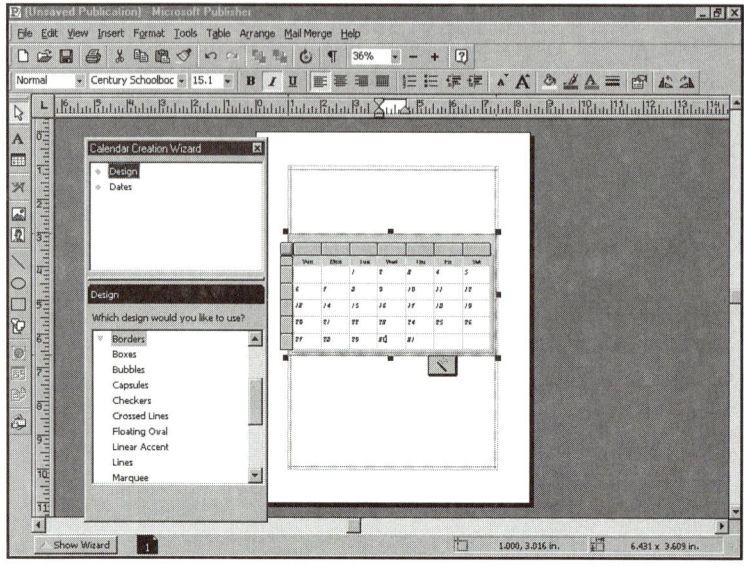

FIGURE 12.2 Each Design Gallery object can be quickly edited using its own wizard.

4. Select a new design (such as Borders in Figure 12.2) in the lower pane of the current Wizard. You can preview as many designs as you want by selecting them in the lower wizard pane.

5. When you finish working with the wizard for a particular smart object, you can close the wizard to give you more working room in the Publisher window. Click the Close (×) button in the upper right corner of the wizard window.

The number of attributes that you can change for a particular Design Gallery object depends on the object. For instance, a calendar enables you to change the overall design and the way dates are displayed. Logos enable you to change the picture associated with the logo, the number of lines in the logo, and the overall design of the logo.

INSERTING OBJECTS FROM OTHER APPLICATIONS

You can also place objects created in other applications into your publication. For instance, you can place an Excel worksheet and a chart into a publication, or you can place a PowerPoint slide on the page of a publication. These types of objects are often referred to as Object Linking and Embedding (OLE) objects.

To insert an OLE object on a publication page, follow these steps:

1. Create an object in any Windows application, such as a worksheet and chart in Microsoft Excel (Figure 12.3). Save the item as a file. Close the application.

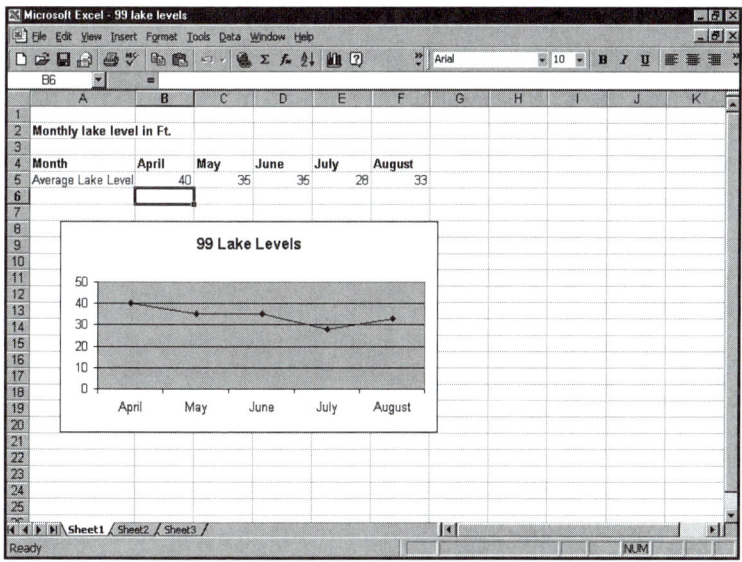

FIGURE 12.3   Any file you create in a Windows application, such as Microsoft Excel, can be inserted into Publisher.

2. In Publisher, click the Insert menu, and then click Object.

3. In the Insert Object dialog box, click the Create from File option button.

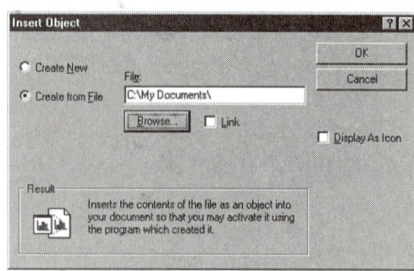

FIGURE 12.4   The Insert Object dialog enables you to insert a new object or an existing object.

4. Click the Browse button to open the Browse dialog box.
5. Locate the file you want to insert as an object, and then select it in the Browse dialog box.
6. Click Insert to close the Browse dialog box and return to the Insert Object dialog box.
7. Click OK to insert the object into Publisher.

> **Creating an OLE Object On-the-Fly**   You can also choose to create your OLE object from scratch during the process to insert into Publisher. From the Insert Object dialog box, choose Create New, select an application from the Object type list, and click OK. Place the new object in the appropriate place on the page. Double-click on the object to edit it.

When you place OLE objects on your publication pages, you can actually edit them using the features of the application that you created them in (such as Excel). Double-clicking on an OLE object in Publisher changes the Publisher menu system and toolbars to provide the features that are available in the server application.

To edit an object, double-click on the object. Figure 12.5 shows an Excel worksheet that is inserted as an object. After the object is activated (by double-clicking), you can edit it as if you were in the server application that the file was created in (in this case, Excel).

 Server Application The application that is used to create the file that is inserted into Publisher as an object.

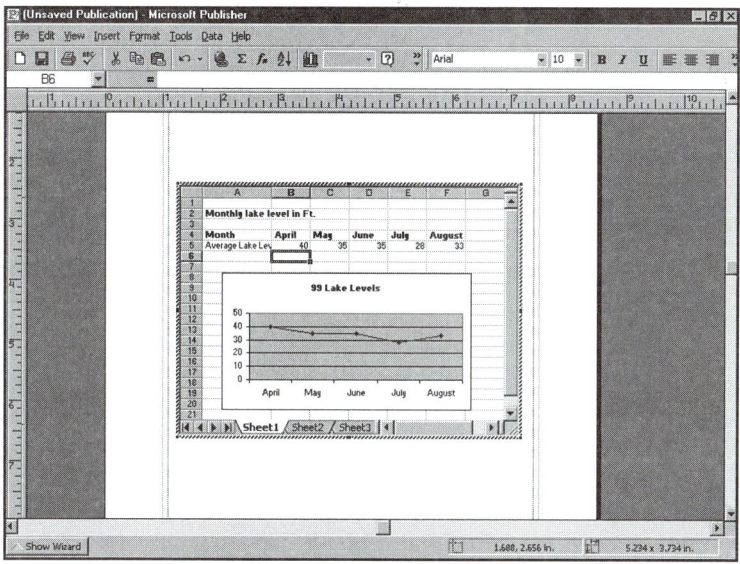

FIGURE **12.5** Double-click an OLE object to edit it.

INSERTING VIDEO AND AUDIO

If you are creating a publication to be viewed on your computer or online, such as a Web site, you can add video and sound to the publication (see Lesson 21, "Creating a Publisher Web Site," for information on using Publisher to build a Web site). This provides you the capability to add some very interesting visual and sound effects to any publication.

Publisher actually provides a library of video animation and sounds that can be inserted into your publications. Both of these media types can be found in Publisher's Clip Gallery.

If you have a sound card and a microphone, you can record your own audio and place it in your publications. Use the Windows sound recorder application to create audio files. You can also insert your own video files if you have video capabilities (a connection for your camcorder or a digital video camera) on your computer.

There are also a number of Web sites that have audio and video clips that you can use in your publications and on your Web site. Use any of the Web search engines (such as WebCrawler) to search for audio and video clips.

Check the Copyright Make sure that clips you use are not copyrighted and that you have permission to use them. Not everything on the Web is in the public domain.

To insert an audio or video clip from the Clip Gallery, follow these steps:

1. To insert an audio or video clip into a presentation, click the Insert menu, select Picture, and then click Clip Art.
2. In the Clip Art window, click the Audio or Video tab.
3. Select a category of sounds or videos (such as Animals on the Video tab) from the selected tab (the categories available depend on the tab you select).

Sample the Clip If you want to play a video or sound clip before you place it in your publication, click the video or sound clip in the Clip Gallery, and then select Play Video or Play Sound from the shortcut bar that appears.

4. Click a sound or video in the window and click the Insert button to place it on the publication page (see Figure 12.6).

ADDING SPECIAL OBJECTS TO YOUR PUBLICATION   115

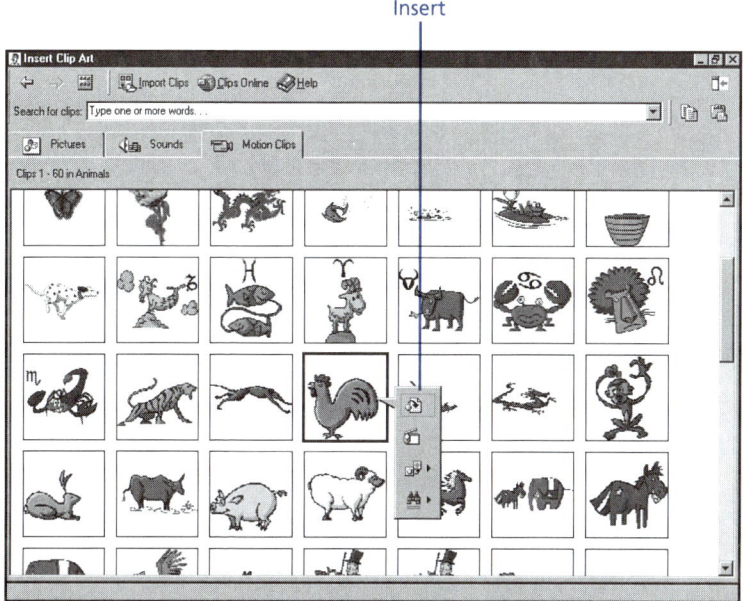

FIGURE 12.6   Insert audio or video clips from the Publisher Clip Gallery.

5. Click the Close (×) button on the upper right corner of the Clip window to close it.

When you have a video or audio clip on the publication page, you can test it by double-clicking. Audio clips play automatically (you hear them immediately). Video clips (depending on the source) open a player application (usually Windows Media Player) that plays the clip. Be advised, however, that the animated clips that you place in your publications from the Clip Gallery only work in publications that have been created as a Web page. The clip plays when you view the publication in a Web browser such as Microsoft Internet Explorer (for more about Web pages, see Lesson 21).

ACQUIRING IMAGES FROM SCANNERS AND OTHER SOURCES

You can also place images into your publications that do not currently exist as a picture or clip art file. You can actually attach to a particular input device and then have it scan an image or take a picture on-the-fly that is then inserted into your publication.

 Input Device Any device such as a scanner or a digital camera that is attached to your computer and can be used to acquire a picture.

Publisher enables you to acquire images from an attached scanner, digital camera, or other device such as a video camera. All you have to do is have the device set up so that it works on your computer and Publisher has no problem using it to capture a particular image.

To capture an image from an attached device, follow these steps:

1. Click the Picture Frame tool on the Publisher toolbar.
2. Place the mouse pointer on the page and drag to create the Picture frame.
3. With the Picture frame selected, click the Insert menu, and then select Picture. Click Acquire Image on the cascading menu. The image currently on your digital camera or on your scanner appears in the Capture dialog box (Figure 12.7).
4. Click Capture to place the image in the Picture frame.

You can size the image just as you do any other frame in your publication. You can also add borders or change the color of the image (for more about enhancing frames, see Lesson 9, "Enhancing Frames with Borders and Colors").

ADDING SPECIAL OBJECTS TO YOUR PUBLICATION   117

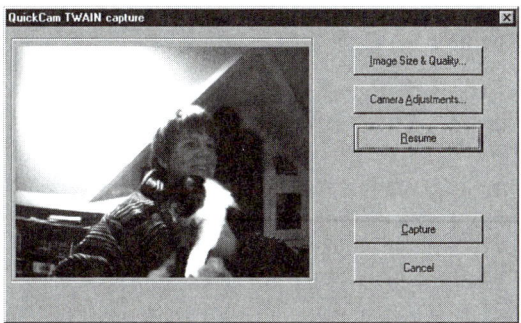

FIGURE 12.7   Images can be acquired from any scanner or camera you have attached to your computer.

In this lesson, you learned to insert special objects using the Design Gallery. You also learned to insert an object from another application and to insert video, audio, and pictures from scanners and cameras into your publications. In the next lesson, you learn how to draw your own objects using the Publisher drawing tools.

Lesson 13
Drawing Objects in Publisher

In this lesson, you learn how to draw objects for your publications using the Publisher drawing tools. You also work with Microsoft Draw to add drawn objects to your publication pages.

Using the Drawing Tools

Although you can insert many different kinds of objects using the Design Gallery or other applications (see Lesson 12, "Adding Special Objects to Your Publications," for more about the Design Gallery), you can also choose to draw your own objects. Publisher provides you with drawing tools that make it easy to draw circles, rectangles, lines, and other shapes. Drawn objects function much the same as any objects you insert into your publications. You can change their fill colors, resize them, and move them on the page.

Publisher provides you with several different drawing tools, which are on the Publisher toolbar. Table 13.1 shows these tools and lists the objects they help you create.

Table 13.1 The Publisher Drawing Tools

Tool	Object
◹	Line
◯	Oval

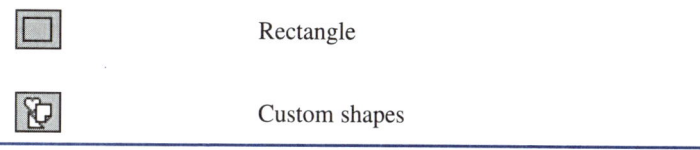

	Rectangle
	Custom shapes

The Line, Oval, and Rectangle tools all function in the same way. You click the appropriate tool and then drag to create the object. The Custom Shapes tool provides a palette of different shapes. Select a particular shape and then drag to create it on the page.

DRAWING A LINE

Publisher makes it easy for you to create a vertical, horizontal, or otherwise-oriented line. Lines that you create using the Line tool on the Publisher toolbar can also include arrow heads and be designed with different line weights and colors.

1. Click the Line tool, and place the mouse pointer on the page.
2. Drag on the publication page to create the line (either dragging vertically or horizontally to create the appropriate orientation) as shown in Figure 13.1.

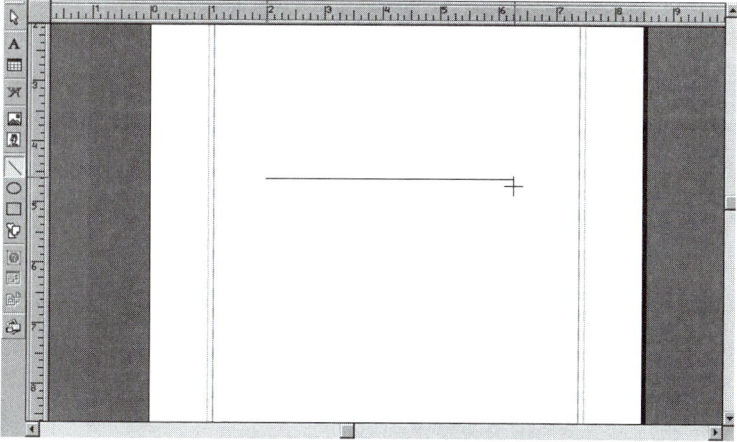

FIGURE 13.1   Drag to create a line on the publication page.

3. Release the left mouse button, and the line appears on the page.

Drawing a Straight Line To draw a perfectly straight horizontal or vertical line, hold down the **Shift** key as you drag to create the line.

To format a line that you draw, double-click the line. The Line dialog box appears. You can control the line thickness, whether the line has arrowheads, and the color of the line in this dialog box.

When you change the formatting options for a particular line, the changes can be viewed in the Sample box of the Line dialog box (see Figure 13.2). After you make all your line formatting selections for the selected line, click OK to close the dialog box.

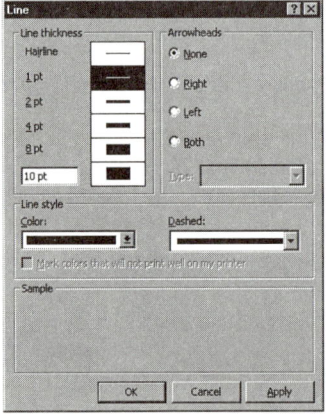

FIGURE 13.2 The Line dialog box enables you to control the formatting for the selected line.

DRAWING AN OVAL OR RECTANGLE

The Oval and Rectangle tools function similarly. You drag in two different directions (down and then to the right, for instance) to create the two dimensions of the object (height and width).

Drawing Objects in Publisher   121

 Getting Your Dragging Technique Down When creating a new drawing object such as a circle or rectangle, drag the mouse downward on the page to create the height for the object, and then drag to the right to create the width for the object. This gives you greater control over the final size of the object. Just don't let go of the left mouse button until you have the object sized appropriately.

To create an Oval or Rectangle, follow these steps:

1. Click either the Oval or Rectangle tool on the Publisher toolbar.
2. Drag on the publication page to create the drawn object (such as a circle).
3. Release the left mouse button, and your new object appears on the page.

 Drawing a Perfect Circle or Square If you want to draw a perfect circle or square using the Oval or Rectangle tool respectively, hold down the **Shift** key as you drag to create your new object.

Drawing a Custom Shape

Publisher also provides you with the capability to draw a number of custom shapes. These shapes range from triangles to stars to sunbursts.

To draw a custom shape, follow these steps:

1. Click the Custom Shapes tool on the Publisher toolbar.
2. Select a custom shape from the shape palette that appears (see Figure 13.3).
3. Drag on the publication page to create the drawn object.
4. Release the left mouse button and your new object appears on the page.

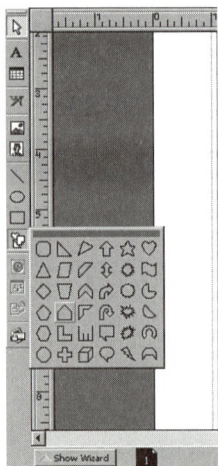

FIGURE 13.3 Select from a number of custom shapes.

Custom shapes can be formatted the same as circles or rectangles created using the drawing tools. The next section discusses formatting the borders and interior color of a drawn object.

> **Layering Objects** You can layer drawn objects using the commands on the Arrange menu. Use the same techniques discussed in Lesson 8, "Working with Publication Frames," under the heading "Arranging Frames in Layers."

FORMATTING DRAWING OBJECTS

You can change the outside border color and the interior fill color of any drawn object. Both of these color attributes can easily be changed on a selected object by using the Line Color and Fill Color buttons on the formatting toolbar.

To change the border color of a drawn object, follow these steps:

1. Click the drawn object that you want to change the line color for.

 2. Click the Line Color button on the formatting toolbar.

3. Select a color on the color palette that appears.

Changing the fill color is a similar operation. Follow these steps:

1. Select the object that you want to select a fill color for.

2. Click the Fill Color button on the formatting toolbar.

3. Click a fill color on the Color palette that appears. If you want to choose from more colors than provided by the Fill Color menu, click the More Colors button on the Color palette. The Colors dialog box appears. Click a color box to select a new color for your object border. Then click OK to close the dialog box. You return to the Color palette and can make your selection.

Rotating an Object

You can easily rotate and flip objects that you create on your publication pages. Rotating an object enables you to reorient an object to the right or the left. For instance, if you want to make a horizontal line a vertical line, you can rotate the line to the right.

To rotate an object, follow these steps:

1. Click the object you want to rotate to select it.

2. Click the Rotate Right (or Rotate Left) button to rotate the object in that direction. The object rotates.

The Rotate Right and Rotate Left commands rotate an object 90 degrees in the either the right or left direction respectively. If you want to custom rotate an object, select the object, and then click the Custom Rotate button on the standard toolbar. In the Custom Rotate dialog box that appears, click either the Rotate Left or Rotate Right buttons to increase the rotation angle in that direction.

You can also change the orientation of an object by flipping it. For instance, you might have a right-pointing object (such as a clip art arrow) that you want to orient so that it points to the left. You can flip the image to the left to get a mirror image of its original orientation.

To flip an object, follow these steps:

1. Click the object that you want to flip to select it.

2. Click the Flip Horizontal button to reverse the orientation of the object horizontally (or click Flip Vertical to flip the object vertically).

DRAWING WITH MICROSOFT DRAW

You can also choose to draw your own images using Microsoft Draw. Microsoft Draw is an add-on application that comes with Publisher 2000 and Microsoft Office 2000. It provides an assortment of drawing tools that you can use to create your own images.

The tools available operate a great deal like the Line, Oval, and Rectangle tools on the Publisher toolbar. The difference between the Publisher drawing tools and Microsoft Draw is that the Publisher drawing tools create individual objects, whereas your Microsoft Draw image (made up of a number of drawn elements) is treated as a single object. Table 13.2 shows and describes a number of the tools available.

Table 13.3 The Microsoft Draw Tools

TOOL	OBJECT
	Line
	Arrow
	Rectangle
	Oval
	Text Box
	Fill Color

Drawing Objects in Publisher 125

Line Color

Font Color

When you start Microsoft Draw, the application actually takes over the Publisher window. All the toolbars and commands that are normally found in the Publisher window are replaced by toolbars for Microsoft Draw. When you click outside the Drawing frame, you are returned to Publisher and its toolbars.

1. Click the Insert menu, and then select Picture. Click New Drawing on the cascading menu. The Draw toolbar appears.
2. Click a tool on the Draw toolbar (such as the Rectangle), and drag to create an item in the Draw frame.
3. Use other tools on the Draw toolbar as needed to complete your drawing (see Figure 13.4).

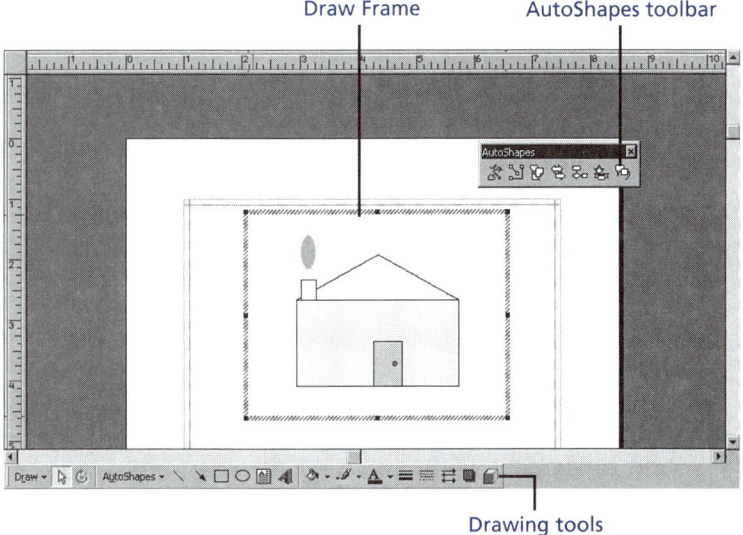

FIGURE 13.4 Microsoft Draw provides a number of tools you can use to create your own objects.

4. Click outside the Draw Frame to return to Publisher.

Microsoft Draw provides a number of object shapes and special objects (such as callouts) that enable you to draw some fairly sophisticated objects. Remember, each Draw object is treated as a single object in Publisher, no matter how many individual shapes were used to create the object itself.

Microsoft Draw Provides More Tools Microsoft Draw offers a larger number of tools and shapes than found on the Publisher toolbar. Custom shapes, such as flow chart shapes, block arrows, and star and banner shapes, can be quickly drawn using the AutoShapes toolbar that appears in the Draw window.

In this lesson, you learned to draw and format objects using the Publisher drawing tools. You also worked with Microsoft Draw to create a drawn object. In the next lesson, you learn how to work with line spacing, tabs, indents, and numbered and bulleted lists.

Lesson 14

Working with Line Spacing, Indents, and Lists

In this lesson, you learn how change line spacing, indent text, and format lists with bullets or numbers.

Setting Line Spacing in a Text Frame

Text is added to your publications in text frames (as discussed in Lesson 10, "Changing How Text Looks"). Each text frame can be formatted differently in respect to line spacing, text indents, tab settings, and how lists are formatted (bulleted or numbered). In fact, every paragraph in a text frame can be set up with different attributes.

Publisher considers any text line or lines followed by a line break (when you press the Enter key) a separate paragraph. This means that every block of text followed by a line break can potentially have different line spacing.

To change the line spacing for text in a frame, follow these steps:

1. Place the insertion point in the paragraph you want to change the line spacing for.

2. Click Format and then Line Spacing. The Line Spacing dialog box appears (see Figure 14.1).

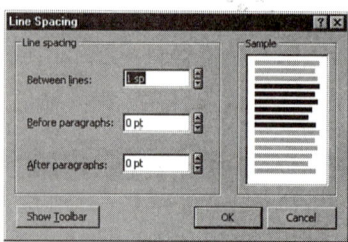

FIGURE 14.1   The Line Spacing dialog box gives you control over the line spacing in a text frame.

3. Type in the new line spacing in the Between lines box (or use the click arrows to select a new spacing).

4. Click OK to close the dialog box and apply the new line spacing.

>
> **Setting Line Spacing Before and After Paragraphs**
> The Line Spacing dialog box also provides you with the capability to add additional spacing before and after paragraphs that appear in a text frame. This makes it easy to offset paragraphs of information from each other or other text elements, such as headings.

After you set the line spacing for a particular paragraph, you can then move to the next paragraph and set a different spacing. In cases where you have several separate paragraphs in the same text frame and want to assign them the same line spacing, select all the paragraphs, and then follow the previously outlined steps.

INDENTING TEXT

You can offset text lines from other text in a text frame by using the indent buttons on the formatting toolbar. Each time you click the Increase Indent button, your text is indented a half-inch from the left edge of the text frame.

Indents work nicely for lists that are grouped under a particular heading. To indent text in a text frame, follow these steps:

1. Place the insertion point in the text line or paragraph you want to indent.

2. Click the Increase Indent button on the formatting toolbar.

You can also decrease the indent that you've placed on a paragraph or particular line of text. Place the insertion point in the line or paragraph, and click the Decrease Indent button.

To indent lines of text that are not part of the same paragraph (lines or blocks of text separated by a line return), select the lines, and then click the Increase Indent button.

SETTING TABS

Another way to indent and align text in a text frame is by using tabs. Tabs are set every half-inch by default. Every time you press the **Tab** key on the keyboard, you offset the text line from the left margin of your text frame by one tab stop. Setting tabs is done by clicking the horizontal ruler to place a particular tab type at a point on the ruler.

You can set different tab types by clicking the Tab selector (it is where the two rulers join in the upper left corner of the screen. The tab types are as follows:

- **Left Tab**   Aligns the beginning of the text line at the tab stop

- **Center Tab**   Centers the text line at the tab stop

- **Right Tab**   Right justifies the text line at the tab stop

- **Decimal Tab**   Lines up numerical entries at their decimal point

130 LESSON 14

Setting a series of tabs on the ruler can enable you to line up text in columns in a text frame. Each tab stop designates the start of a column of text. This is useful when you want to place text in rows of information (for another way to arrange text in rows and columns, see Lesson 15, "Working with Publication Tables").

To set a tab on the ruler, follow these steps:

1. Click the Tab selector to select the type of tab you want to place.

2. Click the horizontal ruler at the position where you want to place the new tab.

3. Repeat steps 1 and 2 to place other tabs on the ruler.

4. Press the **Tab** key on the keyboard, and type the text you want to align at the tab stop (see Figure 14.2).

5. Press the **Tab** key as needed to align the new text at the tab stops.

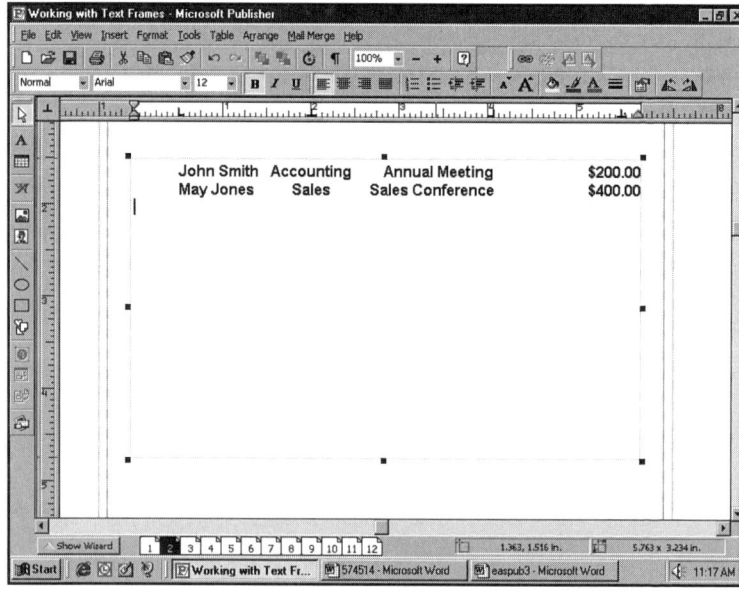

FIGURE 14.2   Tabs can be used to align text in a text frame.

Working with Numbered Lists

You can easily create a numbered list in a Publisher text frame. You have the choice of creating the list before assigning the numbers or turning on the automatic numbering before creating the list from scratch.

The great thing about using Publisher's automatic numbering is that if you add or delete a line in the numbered list, all the numbering changes to accommodate the addition or deletion.

1. Select the text list you want to number.
2. Click the Numbers button on the toolbar. The list is numbered (see Figure 14.3).

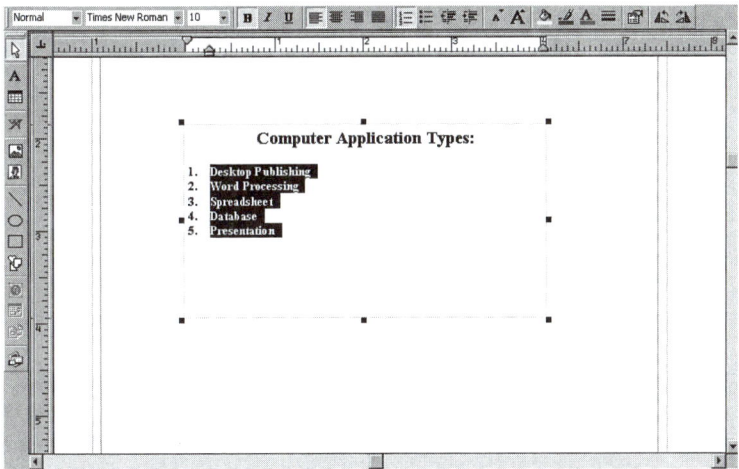

FIGURE 14.3 Lists can easily be numbered using automatic numbering.

3. Click anywhere on the page to deselect the list.

You also have control over the format of the numbering for the list. You can choose the style of the numbering and where the numbers in the list begin (the start value).

To change the number format for a list, follow these steps:

1. Select the numbered list.
2. Select the Format menu, and then click Indents and Lists. The Indents and Lists dialog box opens (see Figure 14.4).

FIGURE 14.4 Change the numbering style for a list using the Indents and Lists dialog box.

3. Use the Format drop-down list to select a new numbering format for the selected list.
4. If you want to change the starting number in the list, type a new value in the Start at box.
5. When you complete your choices, click OK to close the dialog box.

If you want to center or otherwise align the list (rather than using the default indent), click the Alignment drop-down box in the Indents and Lists dialog box, and select a new alignment.

Adding Bullets to Your Text Lists

You can also quickly add bullets to a text list. Bullets help to delineate items in a list from other text in the text frame. You can add bullets to existing lists or turn on the bulleting and then type a new list.

To add bullets to an existing list, follow these steps:

1. Select the text list you want to add bullets to.
2. Click the Bullets button on the toolbar.
3. Click anywhere to deselect the list.

Bulleted lists can also be formatted using the Indents and Lists dialog box. Select the bulleted list, select Format, and then select Indents and Lists.

In the Indents and Lists dialog box (see Figure 14.5) click to select a new bullet type for your list. After you make your selection, click OK to close the dialog box.

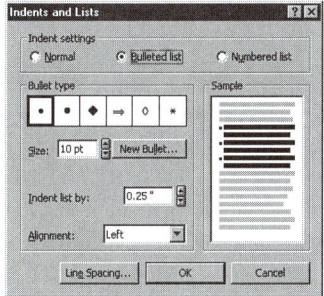

FIGURE 14.5 A number of different bullet types are provided for your bulleted lists.

If you don't see a bullet style that you want to use for your list in the Indents and Lists dialog box, click the New Bullet button. The New Bullet dialog box appears, and you can choose from a number of different symbols to use as bullets.

In this lesson, you learned to set line spacing for text and to use indents and tabs to align text in a text frame. You also learned to add numbers and bullets automatically to lists. In the next lesson you learn to insert and format tables onto your publication pages.

Lesson 15

Working with Publication Tables

In this lesson, you learn to insert a table onto a page. You also learn to move, size, and format the table.

Inserting a Table

Another way to present information in a publication is to use a table. A table enables you to place information in rows and columns, making it very easy to arrange information in a highly accessible format. The intersection of a row and column is called a *cell*. The cells are where you place your data. Publisher gives you complete control over the number of rows and columns in your table and their size. Tables are added to a publication page much the same as any object (such as a text frame or picture frame). You use the Table tool on the Publisher toolbar.

To insert a table onto a page, follow these steps:

1. Click the Table tool on the Publisher toolbar.
2. Click and drag to create the table on the page.
3. In the Create Table dialog box, type the number of rows and columns for the table (see Figure 15.1).
4. Click OK. The table appears on the publication page.

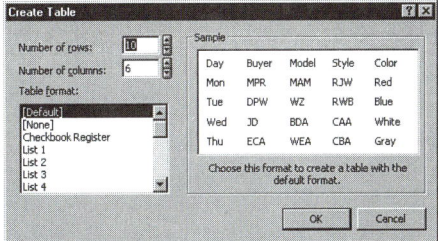

FIGURE 15.1 The Create Table dialog box enables you to select the number of rows and columns for the table.

 Selecting a Format for the Table You can select an AutoFormat for your table in the Create Table dialog box using the Table Format list. For more about auto-formatting, see "Formatting the Table Automatically" later in this lesson.

In cases where you want to place information side-by-side on the page, tables make it a very simple task. Using tables provides you much more control over the placement of information when compared to trying to align items with indents or tabs.

SIZING AND MOVING TABLES

You can resize a table at any time. When you change the table's width (make it larger), this also makes all the column widths larger. If you change the table's height, this changes all the row heights in the table.

To size your table, follow these steps:

1. Click the table to select it.
2. Place the mouse on the appropriate sizing handle and drag to increase or decrease the size (see Figure 15.2).

Lesson 15

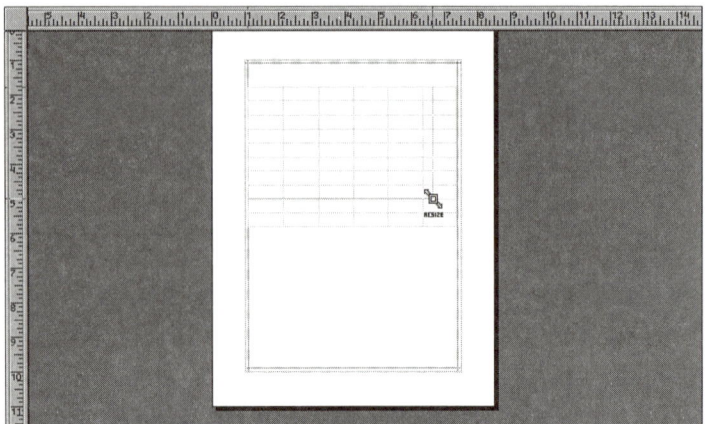

FIGURE 15.2 Size the table as you do any object frame.

3. Release the left mouse button when the table is the appropriate size.

> **Maintaining the Table's Height/Width Ratio** If you want to maintain the table's overall height/width ratio when you increase or decrease the size of the table, make sure to select one of the table's corners and drag on the diagonal.

Because tables are contained in a frame, just like the other objects that you work with in Publisher, you can also easily move them to a new location on the page. You drag them using the mouse.

To change the location of a table, follow these steps:

1. Place the mouse pointer on the edge of the table frame.
2. When the Move icon appears, drag the table to a new location.
3. When the table is in the correct position, release the mouse button.

Paste from One Page to Another You can also copy or cut the table and then paste it on a different page in the publication.

SIZING TABLE COLUMNS AND ROWS

When you change the width of the table, as already shown, it changes the relative width of all the columns in the table. You can also control the individual width of each column in the table to better accommodate text entries that you place in a particular column. This is also true for row heights. You use the mouse to drag a row divider to increase or decrease a row height as needed.

To change a column width or row height, follow these steps:

1. Click the table to select it.

2. Place the mouse pointer on a column divider to the right of the column you want to size (or on a row divider below the row you want to size). The Adjust icon appears.

3. Drag to increase or decrease the width of the column as shown in Figure 15.3 (or drag to change the row height if applicable).

4. Release the mouse button. The column or row is sized accordingly.

 Keeping the Table's Size the Same When Widening Columns If you hold down the **Shift** key while dragging the column width tool, the table stays the same size, and only the column is widened.

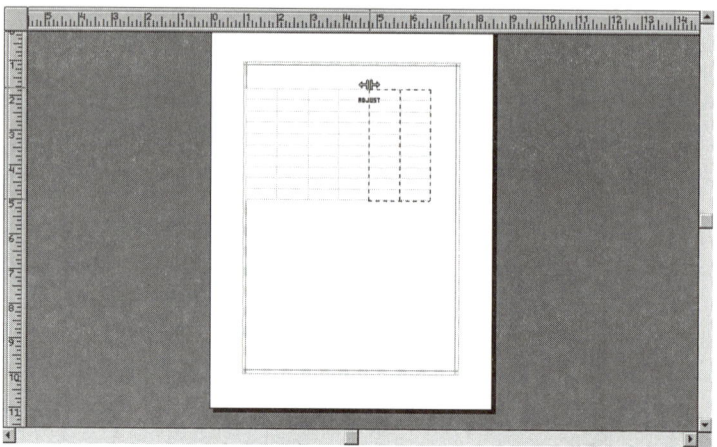

FIGURE 15.3 You can individually size columns and rows.

Using Grow to Fit Text If you are changing row heights in anticipation of typing several lines of text in a row's cell, don't bother. Just make sure that the Grow to Fit Text command is selected on the Table menu. This makes the row's height becomes greater as needed when you type in the text.

ADDING COLUMNS AND ROWS TO THE TABLE

If you find that you need additional columns or rows in the table, you can easily add them. You can add one column or row or add multiple columns and rows if needed.

To add a column or row to a table, follow these steps:

1. Click the table to select it.
2. Click a column or row selector to select it. To add multiple columns or rows, drag to select the appropriate number of column or row selectors (see Figure 15.4).

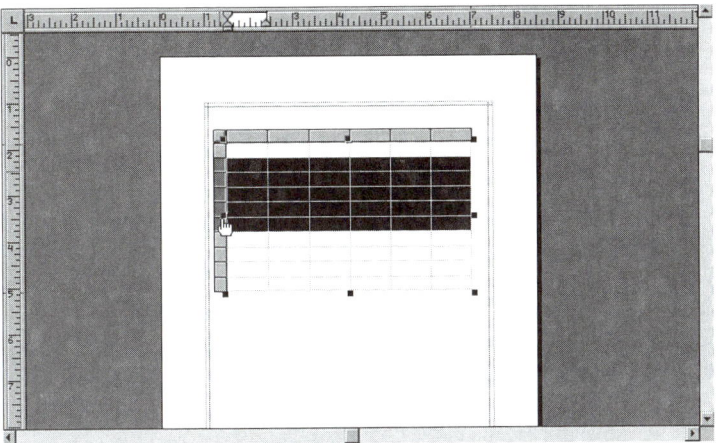

FIGURE 15.4 You can add columns and rows to your table by selecting existing columns or rows.

3. Click the Table menu, and then click Insert Columns or Insert Rows.

The new column (or columns) is inserted to the right of the selected columns, or a new row (or rows) is inserted below the selected rows.

 Resizing the Table After Adding Columns or Rows
Adding columns or rows to a table makes the table and its frame larger as well. You might have to resize tables that you add additional columns to.

You can also use this same technique to delete columns or rows in a table. Select the columns or rows that you want to delete, and then select the Table menu. Then click either Delete Columns or Delete Rows as needed.

USING SPECIAL CELL FORMATS

You also have control over the formatting of the individual cells in your tables. For instance, you can merge several cells (a whole row of cells) and use the merged cells as a place to add a heading for the table.

You can also insert a diagonal line in cells in a Publisher table. This enables you to divide the cell into two areas. This is often used on calendars to divide a cell (a certain day of the week) that holds two dates (such as the 24th and the 31st).

Merging Cells

You might want to take several cells in the same row and make them one continuous cell. The Merge Cells command removes the column dividers between the cells and merges the cells into one continuous cell.

To merge cells in a table, follow these steps:

1. Click the table to select it.
2. Drag to select the cells that you want to merge.
3. Click the Table menu, and then click Merge Cells.

The cells are merged into one cell.

Merging the cells in the first row of a table makes an excellent place to type a heading for the table. Because the entire row is one cell, you can center the heading text using the Center button on the formatting toolbar.

Inserting a Diagonal in a Cell

You can divide a cell diagonally, making the cell in effect two different cells that can contain different information. To divide a cell diagonally, follow these steps:

1. Click the table to select it.
2. Drag to select a cell or cells to split diagonally.
3. Click the Table menu, and then click Cell Diagonals.
4. In the Cell Diagonals dialog box (see Figure 15.5), select Divide down or Divide up.

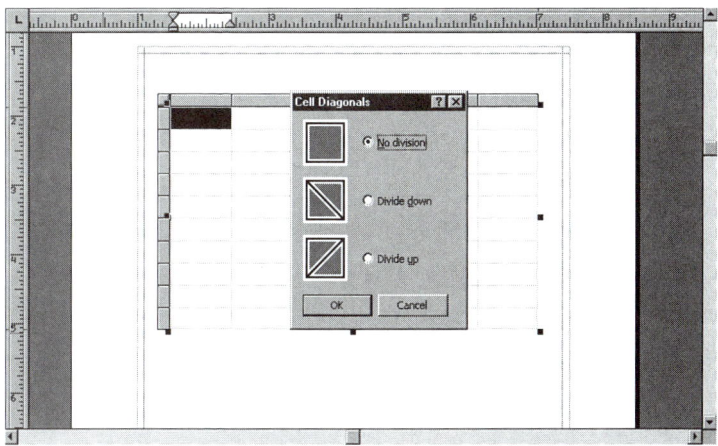

Figure 15.5 You divide a selected cell diagonally using the Cell Diagonals dialog box.

5. Click OK to close the dialog box. The cell is divided by a diagonal line.

To enter text in a cell divided diagonally, click below the line and then enter text at the insertion point. You can click above the diagonal line and enter text in that area as well.

Filling Your Table with Information

When you create a new table, you can quickly enter text in the table cells. After you enter your text, you can format it using any of the text attribute tools discussed in Lesson 10, "Changing How Text Looks."

To enter the text in the table, follow these steps:

1. Click the first table cell you want to enter text in.
2. Type the text for the first cell.
3. Press the **Tab** key to move to the next cell.
4. Enter the text for this cell (see Figure 15.6).

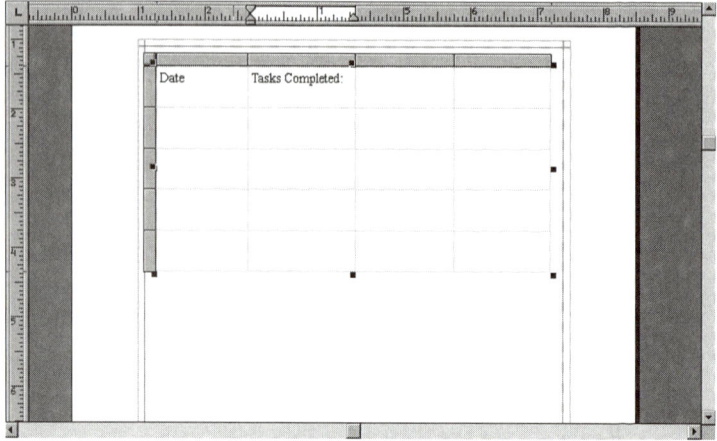

FIGURE 15.6 Enter text in the table cells as you do text in any text frame.

As you type in the table, you can quickly move forward a cell by pressing **Tab**. Press **Shift+Tab** to move back a cell. The up and down arrow keys move you respectively up or down by one cell.

FORMATTING THE TABLE AUTOMATICALLY

You can format a table quickly using one of the AutoFormats provided by Publisher. These AutoFormats provide you with text formatting, cell color formatting, and a number of other formatting attributes. The AutoFormat feature is a great way to create eye-catching tables for your publications.

To automatically format your table, follow these steps:

1. Click the table to select it.
2. Click the Table menu, and then click Table AutoFormat. The Table AutoFormat dialog box appears (see Figure 15.7).
3. Select a format for the table from the list provided.
4. Click OK. The table is formatted.

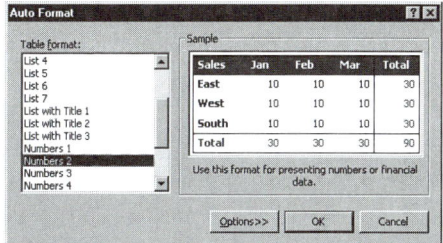

Figure 15.7 Select an AutoFormat to format the entire table.

If you don't like the AutoFormat that you choose, repeat the steps to select a different table format. All the previous formatting is changed to the new AutoFormat.

Formatting the Table Manually

If you want to format the table border and fill colors manually, you can use the Line Color and Fill Color buttons on the formatting toolbar to selectively format the table.

For instance, if you want to add a fill color to certain cells in the table, select those cells (see Figure 15.8), click the Fill Color button, and choose a fill color. You can then add a different fill color to other cells in the table as you want.

Adding border colors to the entire table is even easier. With the entire table selected, click the Line Color button, and choose any of the border colors provided on the Line Color palette.

Formatting tables and cells is like formatting frames. Adding borders to your table or adding borders and fill colors to cells uses the same techniques that you use to enhance frames with borders and colors. See Lesson 9, "Enhancing Frames with Borders and Colors," for more information on working with borders and fill colors.

In this lesson, you learned to insert and format a table. You also learned to automatically format the table and add text to the table. In the next lesson you learn to format publication pages, including their margins, page attributes, and objects placed in the background, such as page numbers.

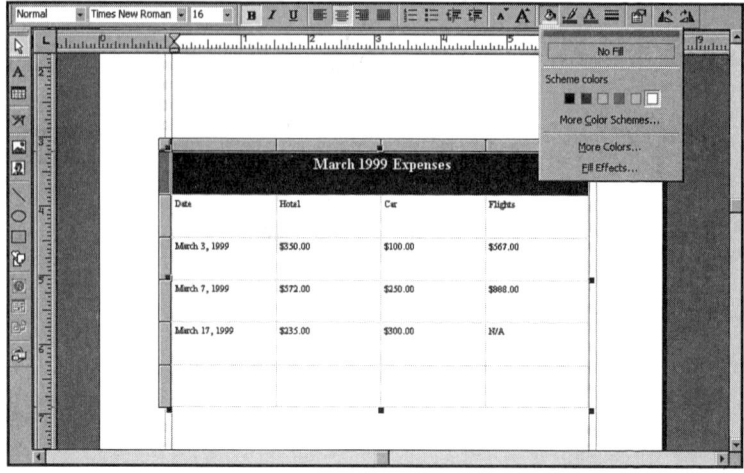

Figure 15.8 Select the cells you want to add a fill color to.

Lesson 16

Formatting Publication Pages

In this lesson, you learn to change the margins on your publication pages. You also learn to add a border to a page, add page numbers to pages, and work in the publication background to insert headers and footers.

Changing Page Margins

When you start a new publication (particularly one that you create from scratch), you might want to change the margins for the publication pages. Publisher enables you to shift the margin guides (the blue and pink lines that surround the page) in the Layout Guides dialog box. You can adjust the top, bottom, left, and right margins.

To change the margins for the current page, follow these steps:

1. Select the Arrange menu, and then select Layout Guides. The Layout Guides dialog box appears (see Figure 16.1).

2. Use the appropriate click arrows to increase (or decrease) the Left, Right, Top, or Bottom margin as needed.

3. Click OK; the new margins appear on your page.

 Don't Change Wizard-Created Publications If you created your publication with a wizard, it's a good idea to leave the margins alone. The wizard placed objects on your pages and set up the margins to accommodate the objects and layout of the publication. Changing the margins might wreak havoc on your page layout.

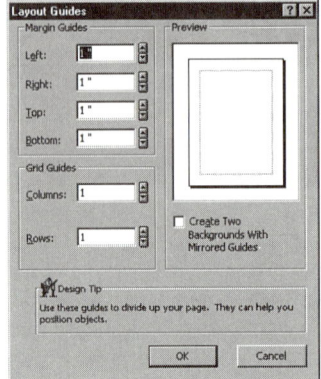

FIGURE 16.1   You can adjust your page margins using the Layout Guides dialog box.

You can also manually drag the margins on a page to a new location with the mouse. However, you must be in the background of the page to manipulate the margins. Select View, and then Go to Background. Hold down the **Shift** key, and drag any margin to a new location. Working in the background of a page is discussed in the section "Working in the Publication Background" later in this lesson.

ADDING PAGE BORDERS

You can place borders around the edge of your pages. This gives you a design element that helps emphasize the objects that are placed within these borders.

When you create a new publication using a wizard or a design set (see Lesson 3, "Creating a New Publication," and 4, "Using Design Sets and Templates," for more information on wizard-created publications and design set publications), the page border for the publication might be part of the design that you select for that particular publication. However, if you want to place a border around pages in a publication that you create from a blank publication template, you can use the Quick Publication Wizard.

To place a border around your publication pages, follow these steps:

1. If the Wizard Pane is not open in the Publisher application window, click the Show Wizard button.

2. Click Design in the top pane of the wizard (see Figure 16.2).

FIGURE 16.2 Borders are in the Design category provided by each wizard.

3. Click to select one of the borders in the Design list in the bottom pane of the wizard pane.

You can try as many of the border designs as you want. Figure 16.3 shows a publication page with a border around the page. When you decide on a particular border, make sure to save your publication.

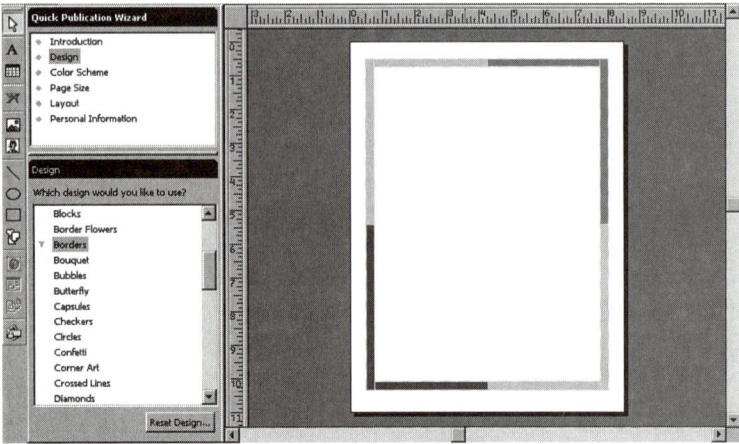

FIGURE 16.3 Borders can add visual interest to the pages of your publications.

 Inserting Blank Publication Pages Negates the Current Wizard If you insert a blank page or pages into a publication that was created using one of the wizards (even a blank publication is created using the Quick Publication Wizard), you can't add a border to these new pages. They are "wizardless" pages. To ensure that the same border is around inserted pages, make sure that the Duplicate All Objects on Page option button is selected in the Insert Page dialog box before inserting the new page. You can delete any duplicated objects that you don't need from the page.

WORKING IN THE PUBLICATION BACKGROUND

When you place the various objects on your publication pages, you are working in the publication foreground. You are placing objects on top of the page. You can also work in the publication background.

Working in the background enables you to easily place repeating elements on every page of a publication. For instance, you might want to number the pages in the publication. You place the page numbering code (discussed in the next section, "Creating Page Numbers in the Background") in the background of the publication, and it shows the appropriate page number on the page.

To go to the publication background, choose the View menu, and then choose Go to Background. All the objects that were on the current page (when you were in the foreground) have disappeared. That is because you are working in a different layer (the background) of the publication.

You can insert a text frame or other object in the background (as in Figure 16.4) that you want to repeat on each of the pages of the publication (as discussed in the next two sections of this lesson). However, you must make sure that objects placed in the background are not obscured by objects that exist in the foreground. For instance, if you place a text frame in the background that holds the current date, it does not show on the publication pages if items in the foreground overlay it.

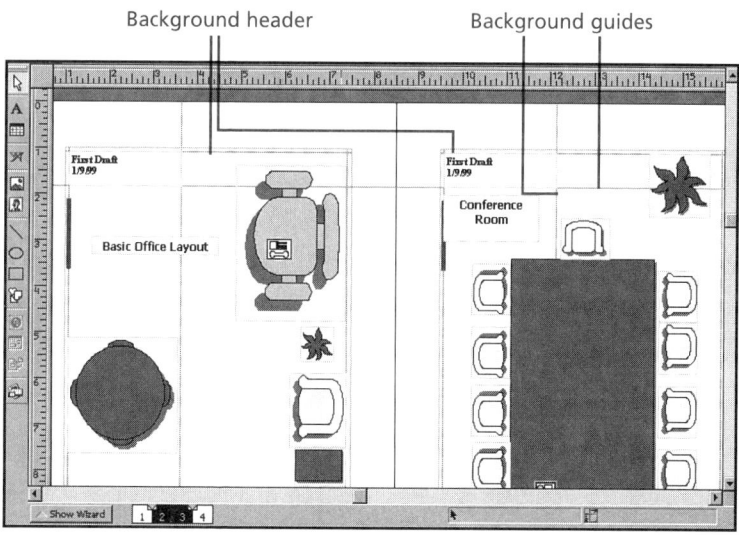

FIGURE 16.4 Placing objects in the background enables you to place repeating elements on your pages.

Figure 16.4 shows a two-page spread of a publication (showing the foreground) where a header text frame has been placed in the background of the publication. Notice that the header appears on both pages. Notice also that guides (that were placed in the background) have been used to make sure the header text frame is not obscured by an object on either of the pages (for more about guides, see Lesson 5, "Viewing Your Publications").

When you finish working in the background of your publication, select the View menu, and then select Go to Foreground to return to the foreground of the current page.

Creating Page Numbers in the Background

When you create publications that contain multiple pages, you might want to number the pages. Page numbers are placed in the background of the publication in a text frame. The great thing about using the page number feature is that even if you insert or delete pages in the publication, the appropriate page number always appears on your pages because Publisher places a page numbering code in the text frame.

To create a page number frame in the background, follow these steps:

1. In the background of a publication page, click the Text Frame tool on the Publisher toolbar.

2. Drag the mouse to create the text frame to hold the page number code.

3. Select the Insert menu and then Page Numbers. A page number code is placed in the text frame (see Figure 16.5).

You might want to have page numbers that read as "Page 1" (and so on). You can click the mouse before the page number code (#) and add the text—"Page" in this example—you want to appear on every page with the page number.

Inserting the Date in a Text Frame

Another useful feature is the capability to automatically insert the date in a text frame. When you place this text frame in the background, you can conveniently place the current date on all the pages of your publication.

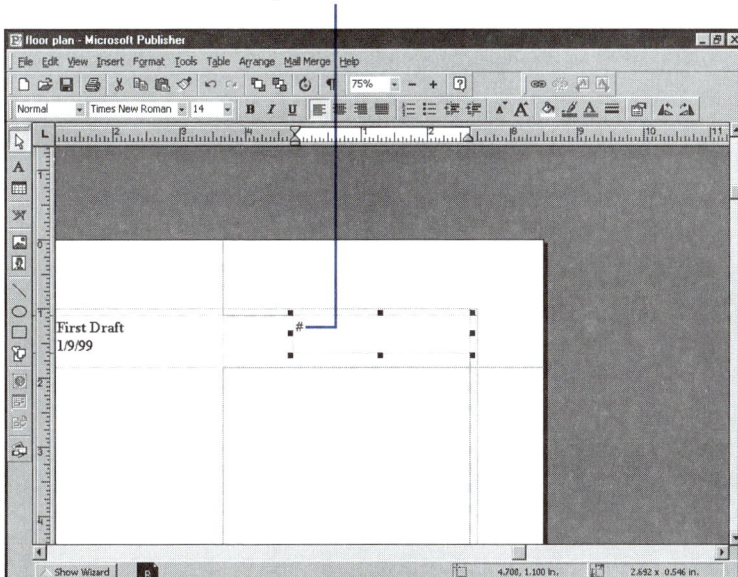

FIGURE 16.5   Place the page number code in the background to number the pages in the publication.

To place the date in the background, follow these steps:

1. In the background of a publication page, click the Text Frame tool on the Publisher toolbar.
2. Drag the mouse to create the text frame to hold the page number code.
3. Select the Insert menu and then Date and Time.
4. Click the date format you want to place in the text frame.
5. Click OK to place the date in the text frame.

Updating the Date Automatically If you want to have the date refreshed (the current date placed into the text frame that you placed the date code into) in your publication whenever you open the saved publication, click the Update automatically check box in the Date and Time dialog box.

In this lesson, you learned to change the page margins and add a border to your pages. You also learned to work in the background and add a page number and date code to a text frame in the publication background. In the next lesson, you learn to fine-tune your publications with the Spell Checker, Design Checker, and AutoCorrect features.

LESSON 17
FINE-TUNING PUBLISHER PUBLICATIONS

In this lesson, you learn to use the spelling feature, control hyphenation in a text box, check your publication design with the Design Checker, and set up the AutoCorrect feature.

USING THE SPELL CHECKER

After you spend a lot of time designing a publication, you might want to print a hard copy of the final product or ready the material for printing by a commercial printing service (Lesson 18, "Printing and Outputting Publisher Publications," covers outputting your publications). Publisher offers several tools that enable you to fine-tune your publication.

The spelling feature checks your documents for misspellings and typos and displays each suspect word in the Check Spelling dialog box. You can choose to replace the word with a suggested correct spelling, ignore the word, or correct the misspelling yourself.

To use the Spell Checker, follow these steps:

1. Select a text frame that you want to spell check.
2. Click Tools, select Spelling, and then click Check Spelling.
3. The Check Spelling dialog box appears (see Figure 17.1). The first suspect word (a word considered misspelled) appears in the Not in Dictionary box. To correct the word, choose one of the following:
    - Select the correct spelling in the Suggestion box, and then click Change.

- To change all occurrences of the word to the new spelling, click Change All.

- To ignore the word (in cases where the word is correctly spelled), click Ignore (click Ignore All to ignore all occurrences of the word).

- If you want to add the word to the dictionary file so that it is not flagged as misspelled in the future, click Add.

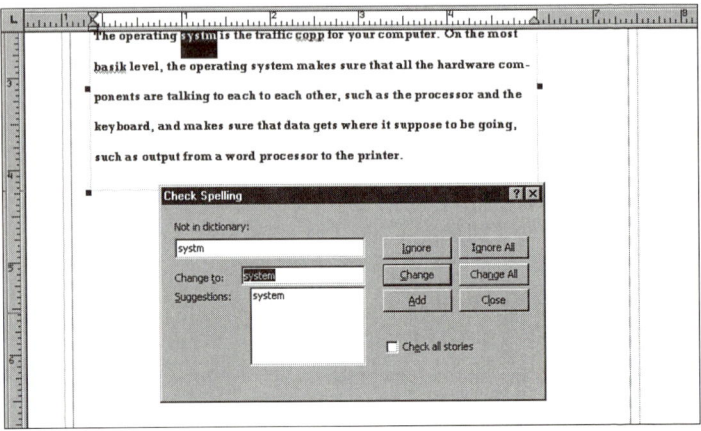

FIGURE 17.1 The Check Spelling dialog box offers you several different options for a flagged word.

4. After you make your selection, the next flagged word is shown in the Not in Dictionary box. Correct other suspected misspellings as needed. When the current text frame is completed, the Spell Checker moves on to the next text box. In situations where you have created more than one story that flow through a series of text boxes (see the section "Connecting Text Frames" in Lesson 10, "Changing How Text Looks"), a dialog box appears asking you if you want to continue checking the publication; click Yes to correct the rest of the publication or No to stop spell-checking with the currently selected text frame.

Check All the Text Frames at Once If you want to automatically check all the text in the publication (all text frames), click the Check all stories check box in the Check Spelling dialog box.

As you might have already noticed, Publisher actually flags suspected misspellings as you type. A red, wavy line is placed under the flagged word. You can choose to correct a flagged misspelling without running the Spell Checker. Right-click the word, and select a correct spelling from the list provided on the shortcut menu that appears.

Controlling Hyphenation in Text Frames

Another element of fine-tuning a publication is determining where words are hyphenated in your text frames. You can have Publisher automatically hyphenate the text in your text frames (which means it determines where to break words with a hyphen and continue the remaining portion of the word on the next line).

When you choose to use the automatic hyphenation feature, hyphens are only placed as needed. The really great thing about the feature is that if you edit the text, unnecessary hyphens are removed automatically, and new hyphens are placed as needed.

To hyphenate a text frame automatically, follow these steps:

1. Click the text box that you want to automatically hyphenate.

2. Click the Tools menu, then select Language, and then click Hyphenation. The Hyphenation dialog box appears (see Figure 17.2).

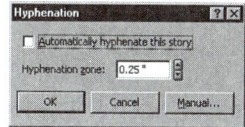

FIGURE 17.2 You can automatically hyphenate text in a text frame.

3. Click the Automatically hyphenate this story checkbox in the Hyphenation dialog box.

4. Click OK to exit the dialog box and hyphenate the text.

You can also choose to hyphenate the text manually. To do this, click the Manual button in the Hyphenation dialog box. A Hyphenate box appears displaying the first word in the text frame that needs to be hyphenated (see Figure 17.3).

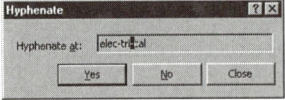

FIGURE 17.3 Manual hyphenation enables you to select the words that are hyphenated.

To use the hyphenation shown in the Hyphenate box, click Yes. If you don't want to hyphenate the word, click No. The next word to be hyphenated is displayed, and you are given the same choices as already described. Continue through the text until you either hyphenate or reject each of the words.

Auto-Hyphenation Might Be On by Default
Depending on your Publisher settings, the automatic hyphenation feature might already be turned on in the Hyphenation dialog box. You can turn it off by clearing the Automatically hyphenate this story checkbox.

USING THE DESIGN CHECKER

The Design Checker is another great tool for helping you fine-tune your publication. The Design Checker actually looks at the design elements and objects in your publication and helps you find empty frames, improperly proportioned pictures, font problems (such as too many), and other design problems. The great thing about the Design Checker is that it offers you help when it identifies a potential design problem.

To use the Design Checker, follow these steps:

1. Select the Tools menu, and then select Design Checker. The Design Checker dialog box appears.

2. In the Design Checker dialog box, specify the number of pages you want to check (see Figure 17.4), and then click OK to start checking the publication.

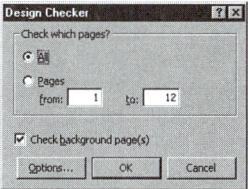

FIGURE 17.4 Use the Design Checker to check your publication for design flaws.

3. When the Design Checker finds a potential design error, you are offered several options (see Figure 17.5):

   - If you want to ignore the flagged design error and move on to the next potential error, click Ignore. Click Ignore All to ignore all suspected errors similar to the currently flagged error.

   - To have the Design Checker fix the current flagged error, click Change. This option is not available in all cases.

   - If the flagged error is an empty text frame, click Delete Frame to remove it.

   - If you're not sure why the Design Checker has flagged the current item, you can get additional help from the Publisher help system related to the problem; click Explain.

   - To move to the next design problem without acting on the currently flagged item, click Continue. When you run the Design Checker again, this problem is flagged again.

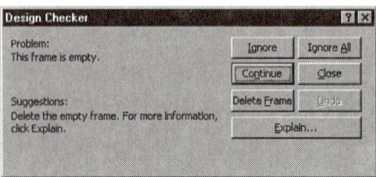

Figure 17.5 You are provided with several different options when checking your publication design with the Design Checker.

4. When you have checked the entire publication (and either ignored or fixed the flagged design problems), the Publisher dialog box appears; click OK to close it.

In many cases, the Design Checker is not able to fix a specific problem for you (as the Spell Checker can) but explains to you what it thinks the potential design flaw is. You can then close the Design Checker and fix the problem. Then, to continue checking the publication, restart the Design Checker from the Tools menu and continue.

SETTING UP AUTOCORRECT

When you add text to your publications, it's nice to have misspellings and typos corrected automatically. The Publisher AutoCorrect feature does this for you. You can set the various AutoCorrect options and add your own common misspellings and typos to the AutoCorrect list.

To set up AutoCorrect, follow these steps:

1. Click the Tools menu, and then click AutoCorrect. The AutoCorrect dialog box appears (see Figure 17.6).

2. To add a common misspelling or typo, type the misspelling in the Replace box and the correct spelling in the With box.

3. Click Add to add the items to the AutoCorrect list.

4. To remove an item from the AutoCorrect list, click the item and then click the Delete button.

5. Click OK when you finish working in the AutoCorrect dialog box.

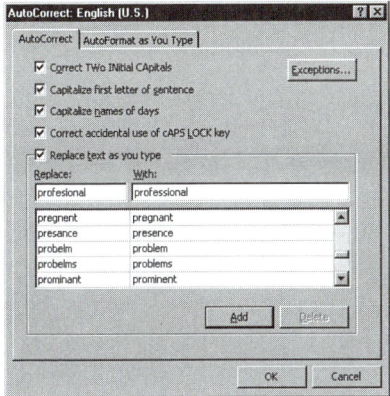

FIGURE 17.6 You can set the options for AutoCorrect in the AutoCorrect dialog box.

The AutoCorrect dialog box also controls some automatic formatting options. To view these options, click the AutoFormat as You Type tab (Figure 17.7).

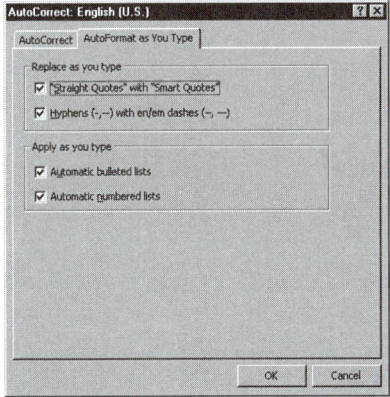

FIGURE 17.7 The AutoFormat as You Type tab controls automatic numbered lists and other formatting options.

This tab controls several formatting options, including the automatic numbering of a list when you start the list with a number and also automatically bulleting list items when you add an asterisk to the beginning of the list. Publisher converts the asterisk to the default bullet type.

All the options on the AutoFormat as You Type tab are on (selected) by default. If you want to not use any of the options, clear the appropriate check box.

In this lesson, you learned to spell-check your publication. You also learned to hyphenate text in text frames, and you checked your publication design with the Design Checker. You also worked with the AutoCorrect feature. In the next lesson you learn to print your publication and ready it for printing by a printer service.

Lesson 18

Printing and Outputting Publisher Publications

In this lesson, you learn to print your publications and ready them for output by a commercial printing service.

Previewing the Publication

When you work in Publisher, you are, in effect, always previewing the publication as it will print. Publisher is a What You See Is What You Get (WYSIWYG) environment. This means that the placement of objects, frames, and other items appear in the Publisher window as they appear on the printed page.

 Web Publications Are the Exception Web site pages you create in Publisher are meant to be viewed in a Web browser. Web pages viewed in the Publisher window and in the Internet Explorer window do differ slightly on the placement of objects and other items.

The best strategy for previewing your publication is to go from the general to the specific. Zoom in and make sure that individual objects are correctly set up and that text boxes do not contain typos. When you zoom out on the publication, you can check placement of objects, the overall design of the publication, and the use of color (for more about using the different views in Publisher, see Lesson 5, "Viewing Your Publications").

Printing the Publication

No matter how hard you work on the color and design parameters of your publication, the final judge of your skills is how the publication appears on the printed page. Publisher actually does a very good job printing both full-color and black and white (grayscale) publications. If you turn on Color Matching, as discussed in the next section of this lesson, you also greatly increase the matching of colors between your monitor and printer.

To print your publication, follow these steps:

1. Click the File menu, and then click Print. The Print dialog box appears (see Figure 18.1).

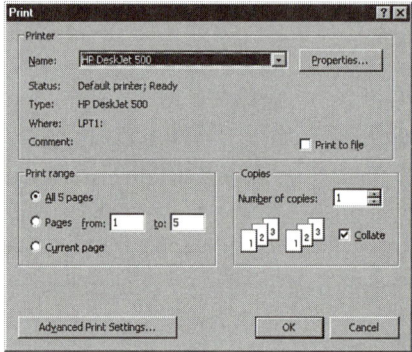

Figure 18.1 Select the page range to print in the Print dialog box.

2. If necessary, click the Name drop-down box, and select the printer you want to send the print job to.

3. Set the page range of the printout if necessary.

4. Use the Number of copies click box to specify the number of copies.

5. Click OK to send the print job to the selected printer.

  Print Quickly with the Print Button You can send your publication directly to the printer and bypass the Print dialog box. Click the Print button on the toolbar.

Your publication prints. If you have problems with the printout or don't like the way it looks, you can work with the various print options to enhance the final printout (discussed in the next section). If you still have problems with the print job, you can use the Print Troubleshooter (see the section "Troubleshooting Printing Problems" in this lesson) to try to solve any printing problems.

Working with Print Options

You can also set a number of different print options before sending the print job to the printer. These options range from such settings as the paper type that is used for the printout and the orientation of the printout (Portrait or Landscape) to more advanced settings, such as Color Matching. Unfortunately, not all these settings are contained in one specific dialog box, so you have to hunt around a little before you find all the print options available.

Setting Paper Type and Orientation

The paper type and page orientation for the printout are set in the Print Setup dialog box. This dialog box looks very much like the Print dialog box. However, even after you set options in the Print Setup dialog box, you still have to return to the Print dialog box to actually print the publication.

To set options in the Print Setup dialog box, follow these steps:

1. Click the File menu, and then click Print Setup. The Print Setup dialog box appears (see Figure 18.2).

2. If necessary, click the Name drop-down box, and select the printer you want to send the print job to.

3. Click the Paper Size drop-down box to select any special paper types you might print to.

4. Click either the Portrait or Landscape option button to select the orientation for the printout.

5. Click OK to close the dialog box.

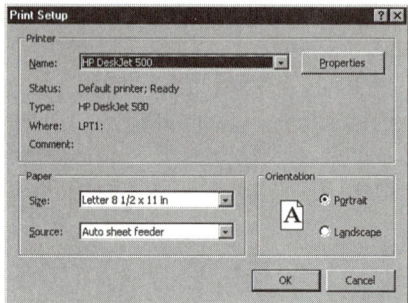

FIGURE 18.2 Set paper type and page orientation for the publication in the Print Setup dialog box.

>
> **Wizard-Created Publications Have Their Own Print Setup Options** If you create a publication using one of the wizards, options such as page orientation and special paper type are set automatically. In most cases, you should not change the Print Setup options for a wizard-created publication.

After you select the various options in the Print Setup dialog box, you can print the publication.

MATCHING MONITOR AND PRINTER COLORS

As you ready a publication for printing, make sure that the colors that are actually printed by the printer are the same as (or at least very close to) the colors that you saw on your monitor as you designed the publication. To aid the printer in matching the colors on the computer monitor, you can turn on the Color Matching feature.

To set up the Color Matching feature, follow these steps:

1. Click the File Menu, and then click Print.
2. In the Print dialog box, click Advanced Print Settings. The Print Settings dialog box appears (see Figure 18.3).

Printing and Outputting Publisher Publications 165

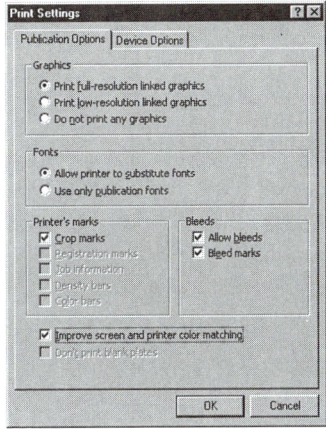

FIGURE 18.3 Color Matching can help you get the colors you see on the screen to the printed page.

3. Click the Improve screen and printer color matching check box.
4. Click OK to close the dialog box.
5. You can send the publication to the printer by clicking OK or click Cancel to close the Print dialog box without printing.

Even if you do turn on the Color Matching feature, you might find that some colors still do not print the same as they appear on your monitor. You can have Publisher mark these colors for you in the color palette, so that you can avoid using them in future publications. Colors that do not match are typically a result of your printer's inability to properly create that color.

Select any object on the page, then click Format, select to Fill Color, and then click More Colors. Click the Mark colors that will not print well on my printer check box.

Colors that do not print well are now displayed with an X in the color palette. For instance, on a black and white printer, all the colors on the palette except for white and black are marked with an X.

TROUBLESHOOTING PRINTING PROBLEMS

Although Publisher makes it fairly simple for you to print your publications to a printer attached to your computer, you might find that upon occasion you have a problem. Your text might not look right, a portion of a picture might be cut off, or the printer might not print the publication at all. When you run into printing problems, you can call on the Print Troubleshooter, which provides you with advice on how you might overcome a particular printing problem.

To use the Print Troubleshooter, follow these steps:

1. Click the Help menu, and then click Print Troubleshooter.

2. The Help window appears on the desktop (see Figure 18.4). Click one of the print troubleshooting topics in the Help window (for example, click My text or fonts do not look right).

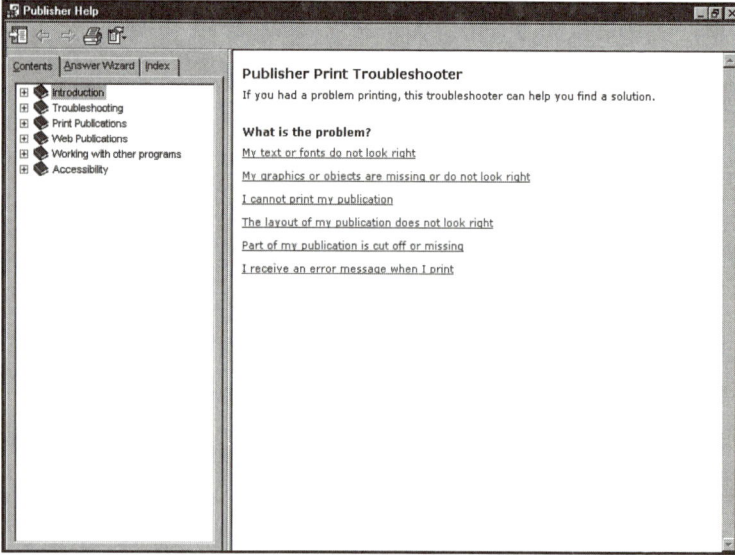

FIGURE 18.4 The Print Troubleshooter can help you fix printing problems.

3. A second, more specific level of help topics associated with your last selection appears. Click a selection to view the help associated with it (such as My text does not wrap correctly).

4. Help is provided to specifically remedy the problem you selected. When you finish reading the remedy for the problem, click the Close button to close the Help window.

If a particular solution provided by the Print Troubleshooter does not alleviate your current problem, you might want to open the Troubleshooter again and try a different solution. If you can't print at all, you might want to make sure your printer is turned on, plugged in, and securely attached to your computer via a printer cable. Physical connections are often the culprit when you can not print at all.

Working with an Outside Print Service

In some cases, you might design a publication that you don't print on your own printer. For instance, you create a flyer or a business card, and you want to have the publication placed on special papers using high quality inks that only a commercial printer can provide. You can use the commercial printing tools in Publisher to ready any publication for printing by a professional printer.

Two of the most important aspects of readying a publication for outside printing are to select the appropriate color scheme (so that it matches the one used by the printer) and to save embedded pictures and graphics as separate files that are linked to your publication. This is a necessity if you use a professional printer.

Setting the Color Scheme

The color scheme that you select for your publication is determined by the print service. All you have to do is use the Color Printing dialog box to specify the particular color scheme.

To set the color scheme for your publication, follow these steps:

1. Click the Tools menu on the Publisher toolbar, then select Commercial Printing Tools, and then click Color Printing. The Color Printing dialog box appears (see Figure 18.5).

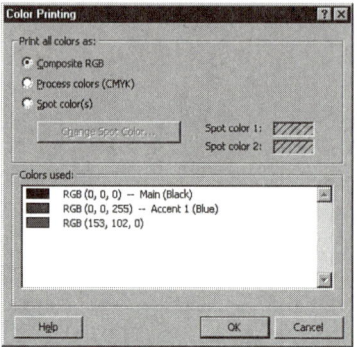

FIGURE 18.5 Select the color system for your printing service to use.

2. In the Color Printing dialog box, click the radio button for the color scheme specified by the printing service that you use:

   • Composite RGB   This is the default color system and is used by most color printers you use at home or in the office. This is the setting to use if you print the publication yourself.

   • Process Colors   CMYK is a color system used by most commercial printers. It works particularly well on publications where color photos are included.

   • Spot Colors   For black and white publications that only use one or two additional colors to emphasize items on the pages, select spot colors.

3. If you select Spot Colors, click the Change Spot Color button to edit the spot color (the color other than black and white that appears in the publication).

4. In the Choose Spot Color dialog box, use the drop-down lists to choose Spot color 1 and Spot color 2 (if one exists) that are to be used on the publication.

5. Click OK to return to the Color Printing dialog box.

6. After you select your color scheme, click OK to close the dialog box.

LINKING GRAPHICS TO THE PUBLICATION

When you place your pictures and clip art on a publication page, you are making that image part of the page. When you have a publication printed by a commercial service, you typically have to provide the pictures and clip art images as separate files that are linked to the publication rather than placed in it. The Commercial Printing Tools provide a Graphics Manager that enables you to convert your publication objects to linked objects before you take the publication to a printing service.

To create separate graphics files for your publication, follow these steps:

1. Click the Tools menu on the Publisher toolbar, then select Commercial Printing Tools, and then click Graphics Manager. The Graphics Manager dialog box opens.

2. In the Graphics Manager, click the first picture or clip art item in your publication (see Figure 18.6), and then click Create Link. The Create Link dialog box appears.

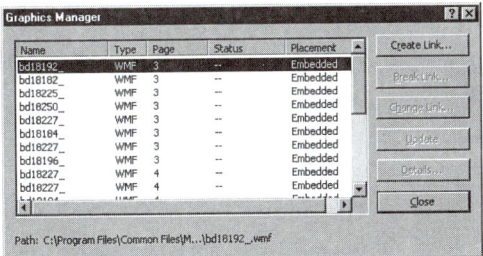

FIGURE 18.6 Select the first picture or clip art to save as a separate file.

3. In the Create Link dialog box, click the Create a file radio button.

4. Click OK to create the linked file. The Save As dialog box appears.

5. Click Save in the Save As dialog box, and the linked file is created (you can go with the default file name for the object).

6. Repeat the preceding steps to create other links as needed for your graphics files.

7. When you finish creating files for all the graphics, click the Close button.

USING PACK AND GO

A publication that is jampacked with pictures, text frames, and various design elements constitutes a file that is a pretty good size. In many cases, the publication might be too large to fit on a diskette. However, you can compress your publications using Pack and Go. This makes it easy for you to take your publication to another computer or take it to a commercial printer.

To pack a presentation, follow these steps:

1. Click the File menu, then select Pack and Go, and then click Take to Another Computer. The Pack and Go Wizard appears.

2. Click Next to start the Pack and Go Wizard.

3. To pack the publication to a disk, click Next (or specify another drive on your computer, and then click Next).

4. To embed the fonts and pictures in the publication in the Pack and Go file, click the appropriate check boxes (see Figure 18.7), and then click Next.

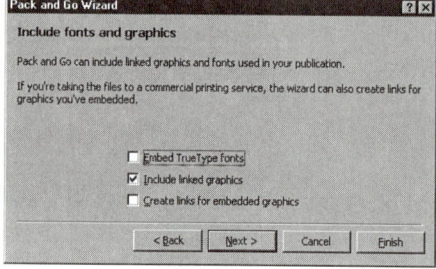

FIGURE 18.7 Select items to include in the Pack and Go file.

5. Click Finish to pack the publication onto a disk.

If the packed publication does not fit on one diskette, the Pack and Go Wizard prompts you to place additional diskettes in your floppy disk drive.

A copy of Unpack.exe is also placed on the diskette (or diskettes) that holds your packed presentation. You can use this program to unpack your publication on another computer.

To unpack a packed file, place the diskette in the other computer, and then click Start and Run. In the Run box, type **a:\unpack.exe**. You are prompted for a destination folder for the unpacking of your publication; specify the folder, and then click OK.

In this lesson, you learned to print your publications. You also learned to set print options and ready your publication for printing by an printing service. In the next lesson, you will learn how to create publications for mass mailings using the Publisher merge feature.

Lesson 19

Mass Mailing Publications

In this lesson, you learn to create a mailing list in Publisher and then merge the list with a publication for mass mailings.

Understanding the Mail Merge Feature

The Publisher merge feature actually helps walk you through the process of creating a publication, such as an envelope or mailing label, that can be churned out in large quantities for mass mailings. The merge process uses the publication and a data source to create a personalized version of the publication for each individual in the data source.

> **Data Source** A data source is a file that contains names and addresses that are to be merged into a publication. Data sources in Publisher are really like miniature database files. Publisher makes it easy for you to create a data source using the Publisher Address List feature.

So, to complete a merge, you only have to create two different items. You need to create a mailing list that contains the names and addresses of the people you want to receive the publication. You also need to create a publication, such as an envelope or mailing label or other publication (using a wizard or a blank design template). When you have these two items, you can start the merge process and insert *merge codes* (derived from the fields in the mailing list) as placeholders into the publication (the envelope or mailing label).

After the merge is complete, you can send the merged publications to your printer. The process is really that straightforward. You can use any type of publication with the merge feature.

 Merge Codes Merge codes are the names of the various fields of information that you place in the mailing list. Publisher automatically sets up field names (merge codes) for you, such as First Name, Last Name, and so on.

BUILDING A MAILING LIST

The first step in creating a merge is to build a mailing list for the publication that is to be mass mailed. After the mailing list is created, it's easy to use the merge feature to personalize the publication for each person on the list.

To create a Publisher Mailing List, follow these steps:

1. Click Mail Merge, and then click Create Publisher Address List. The New Address List dialog box appears. The dialog box provides blank address fields (see Figure 19.1).

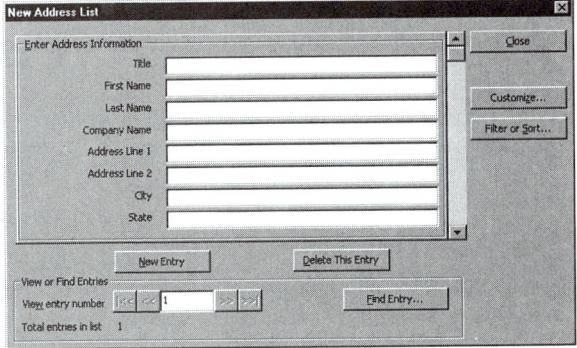

FIGURE 19.1 Fill in the blanks in the New Address List dialog box to build a mailing list.

2. Type the appropriate information for your first addressee in each of the fields in the New Address List (press the **Tab** key to move to the next field).

3. When you have finish entering the information, click New Entry to enter the next person.

4. Repeat steps 2 and 3 as necessary. Click Close when you finish entering all the names and addresses.

When you close the Address List dialog box, you are asked to save your new mailing list (see Figure 19.2). Follow these steps to save the file:

1. Type a name for your mailing list file in the File name box.

2. Select the drive and folder where you want to save the file.

3. Click the Save button.

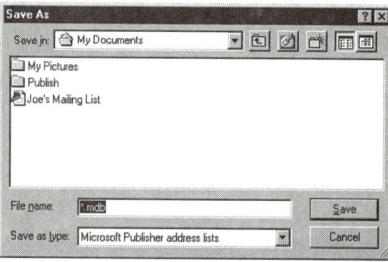

FIGURE 19.2 Type a name and specify a location for your mailing list file.

Use Other Data Sources If you use Microsoft Access and have already created a table of names and addresses for your contacts, you can use it as your mailing list for merges with Publisher publications. Publisher actually saves its mailing list in the Access (.mdb) file format.

After you create the mailing list, you can use it over and over with any number of publications.

You can also edit your mailing list as needed. Select the Mail Merge menu, and then select Edit Publisher Address List. Open your mailing list in the Open dialog box. You can edit entries, add entries, or delete entries from the file.

STARTING THE MERGE AND INSERTING MERGE CODES

After you create a mailing list, you need to place the merge codes in the publication. As already discussed, these codes are placeholders for information that is to be pulled from the mailing list file. For instance, to place the first name of a person in the merged publication, you need to enter the First Name code into the publication before you actually perform the merge.

To have access to the appropriate merge codes, you actually start the merge process and identify the mailing list that the codes are derived from.

To begin the merge process and insert the merge codes, follow these steps:

1. Open the publication (such as a mailing label or envelope), and create a text frame to hold the merge codes (see Figure 19.3).

2. Click Merge and then Open Data Source. The Open Data Source dialog box appears.

3. Click Merge information from another type of file to open a second Open Data Source dialog box (it looks a lot like a typical file open dialog box; see Figure 19.4).

4. Select your mailing list file, and click OK.

   After you open the data source, an Insert Fields dialog box appears. It lists all the fields that are contained in the Mailing List form that you filled out for each of your recipients.

5. Select a merge code in the Insert Fields dialog box (see Figure 19.5).

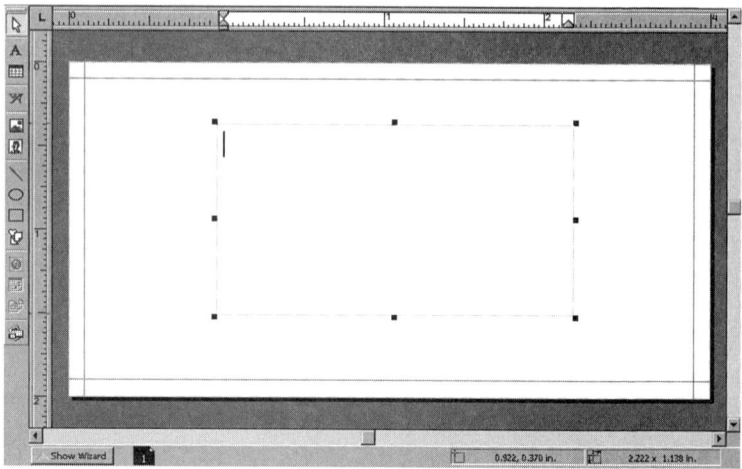

FIGURE 19.3   Create a text frame on the publication to hold the merge codes.

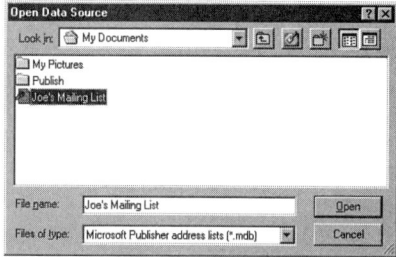

FIGURE 19.4   Specify your Mailing List file in the Open Data Source dialog box.

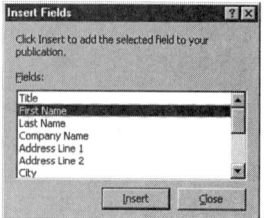

FIGURE 19.5   Select the field in the dialog box, and then click Insert.

6. Click Insert to insert the appropriate merge code.

7. If you need to place a space after the current merge code before you insert the next merge code, drag the dialog box out of the way and click in the text frame. Type a space or other punctuation mark as needed.

8. Insert the next merge code as described in steps 5 and 6. If the next merge code needs to be on a new line, click in the text frame at the end of the first line and press **Enter**.

9. Insert all your merge codes as required. When you have finish inserting the various fields, click Close.

Figure 19.6 shows a text frame on an envelope that contains merge codes.

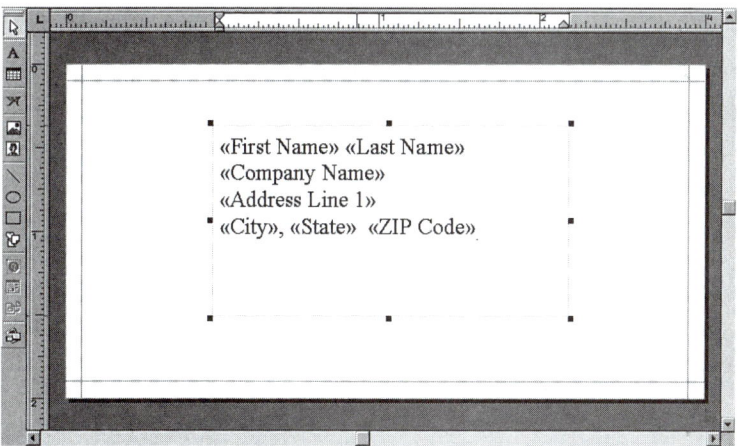

FIGURE 19.6   Insert the necessary merge codes into the text frame.

COMPLETING THE MERGE

After you insert the merge codes into a specific publication (in a text frame), you are ready to perform the merge. As you perform the merge, you are actually given an opportunity to preview the results. You can then send the merged publication to the printer connected to your computer (or your network), and a personalized publication is printed out for each individual in the Mailing List.

To complete the merge, follow these steps:

1. Click the Mail Merge menu, and then click Merge. The data from the first record in your mailing list appears on the publication in place of the merge codes (see Figure 19.7).

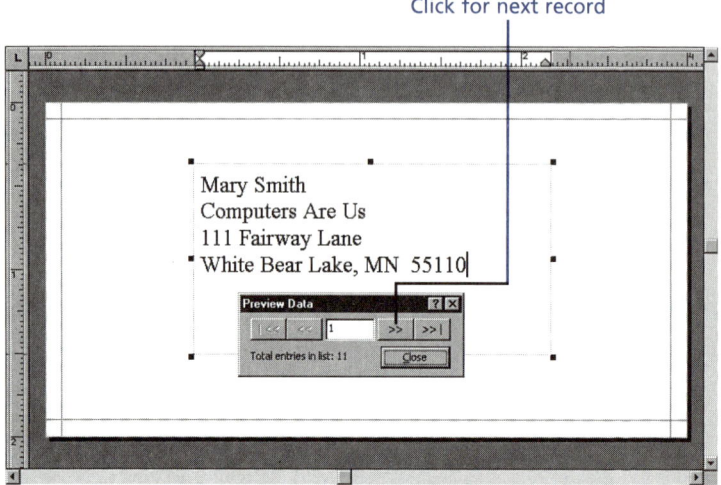

FIGURE 19.7 A preview is given of each record in the mailing list.

2. Click the Next Record button on the Preview Data dialog box to view the next record.
3. Click Close when you finish previewing the merge.
4. To print the merge, select File, and then select Print Merge.
5. Click OK. The merged publications is sent to the printer.

 Stop the Merge! If you find that there is a problem with the merge when you preview your entries, click Mail Merge and then Cancel Merge. You can edit the mailing list or publication and try the merge again.

In this lesson, you learned to create a mailing list and insert merge field codes into a text frame in a publication. You also learned to run a merge and print out the merged publications. In the next lesson, you learn to create publications that require special paper, such as business cards, envelopes, and mailing labels.

Lesson 20

Creating Publications on Special Paper

In this lesson, you learn to create publications that require special papers such as trifold brochures, business cards, envelopes, or mailing labels.

Creating Trifold Brochures

Publisher gives you the capability to create publications that use special papers. These special papers can take the form of scored brochure paper (scored to fold), perforated business card papers, various mailing label sheets, and different envelope sizes. Although many folding publications, such as brochures and greeting cards, can be created on regular printer paper, special papers provide added design elements and ease of use features (such as scoring or fold lines) that regular paper cannot provide. Special papers are provided by a number of different companies, such as Paper Direct, Avery (the label company), and others.

Trifold brochures can normally be one of the most complicated publications to tackle because you must visualize how the paper folds and then place the text and objects on the correct panel of the brochure. Publisher, however, makes it easy for you to create brochures and also takes into account that special paper types exist for your use.

To create a brochure using a special paper, follow these steps:

1. Select the File menu, and then select New. The Publisher Catalog appears.

2. Click Brochures in the Wizards pane of the Catalog.

3. To view the special paper brochures, click Special Paper Informational.

4. To view the different special paper brochures, scroll through the brochures in the Special Paper Informational Brochures pane of the Catalog (see Figure 20.1)

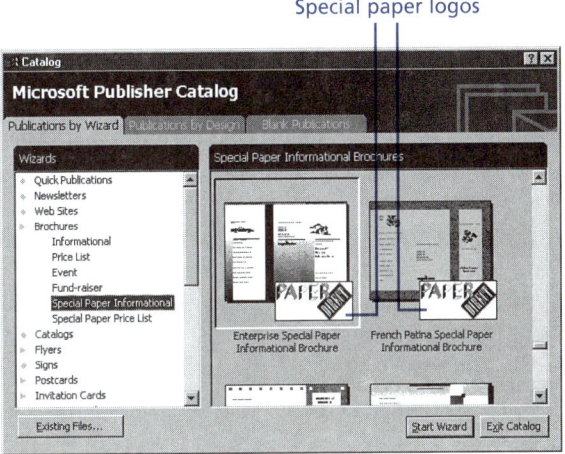

FIGURE 20.1 Special paper brochures use papers that already contain design elements and colors.

5. Special paper brochures are marked by the logo of the paper manufacturer (such as Paper Direct). The special paper name (such as Aristocrat Special Paper) is also noted on each brochure. Click the brochure that you want to create.

6. Click Start Wizard to create the brochure and walk through the wizard steps that enable you to make design and color decisions related to the brochure.

After you create the brochure and are ready to print it, you can load your printer with the special paper. Make sure to read any information that came with the paper that gives you tips on how to orient and load the paper in your specific printer.

Design Decisions Are Limited When you use a special paper, the wizard does not provide you with as many color and design decisions as are available with some plain paper publications. This is because the overall design and layout of the brochure is closely tied to the special paper that you use to print the brochure on.

 Special Papers Can Give Mixed Results When you use very shiny, high-gloss special papers, you might find that your inkjet printer does not give you the same quality of print job that you get on regular paper. In some cases, you might need to buy special papers that are designed for color inkjet printers because the paper must let the spray of ink penetrate the paper surface easily.

CREATING BUSINESS CARDS

Business cards are another publication where using special papers can provide you with a much better result than using regular card stock. A number of companies make business card paper that is of a card stock quality and comes in 8.5"×11" perforated sheets that fit in most printers.

Creating business cards using a wizard is your easiest route to business cards that take advantage of special paper. To create the business card, follow these steps:

1. Select the File menu, and then select New. The Publisher Catalog appears.
2. Click Business Cards in the Wizards pane of the Catalog.
3. To view the special paper brochures, click Special Paper.
4. Select any of the special paper business cards in the Special Paper Business Cards pane of the Catalog, and then click Start Wizard to begin the business card creation process.

Figure 20.2 shows a business card created using the Paper Direct Influential business card paper. The greatest advantage of using special paper for business cards is that the sheet is perforated, so each individual card can be pulled from the main sheet and have fairly straight edges. This enables you to create business cards that can look practically as good as commercially printed cards.

FIGURE 20.2   Special papers enable you to create professional-looking publications.

There are a number of companies that make special papers for your printers (including Hewlett Packard, which makes many different kinds of printers). If the special paper you purchased isn't listed, do a little experimenting. Try to match to up to one of the special papers listed for the type of publication you want to create. Then create the publication, and print a test page. If the paper works, you can mass-produce your work. If it doesn't work, fine-tune the publication until you get an adequate printout.

CREATING ENVELOPES

Envelope creation isn't really an issue of special paper as much as having the right envelope size selected. You can create envelopes using a wizard following the same sequence of steps discussed for special paper brochures are business cards (and special papers are available for envelopes).

If you want to quickly print an address or create an envelope for a mail merge (as discussed in Lesson 19, "Mass Mailing Publications"), you can open a blank publication and then change the page size in the Page Setup dialog box.

184 LESSON 20

To create an envelope from a blank publication, follow these steps:

1. Click the New button on the standard toolbar. A blank publication opens in the Publisher application window.

2. Select the File menu, and then select Page Setup. The Page Setup dialog box appears

3. In the dialog box, click the Envelopes option button (see Figure 20.3).

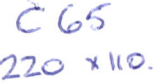

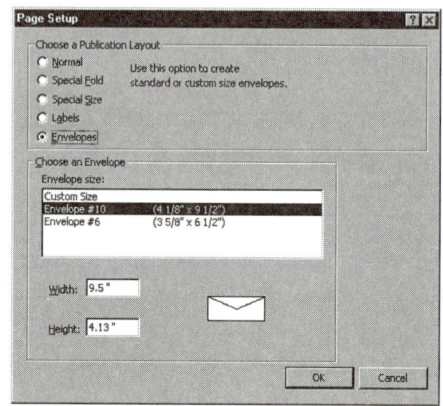

FIGURE 20.3 Select the envelope size in the Page Setup dialog box.

4. Choose the envelope size in the Envelope size box.

5. Click OK to return to the publication.

The publication is now an envelope. Place text frames on the envelope as needed to place an address or field merge codes on the envelope for a mail merge.

When you create an envelope using a wizard or selecting an envelope from a design set, the envelope size is set for you in the Page Setup dialog box.

CREATING MAILING LABELS

The mailing label is another special publication type that deserves some discussion. Many companies make mailing labels for printers. Avery mailing labels are considered the standard, however, and the wizards and Page Setup dialog box both use Avery label numbers to configure the size and number of labels that come on the label sheet that you place in your printer.

If you want to use labels made by a company other than Avery, many manufacturers supply the Avery number that their label is an equivalent to. Look for this information on the label packaging. You can also measure and count the labels that you have on the sheets that you purchased and try to match them up with the specifications for one of the Avery label types listed.

You can create labels by using a wizard or formatting a blank publication as a label using the Page Setup dialog box. If you use a wizard, as was discussed for brochures and business cards, select the publication wizard that uses the Avery number of the labels that you plan to use in your printer.

If you want to quickly set up a label publication with no frills for a mail merge, follow these steps:

1. Click the New button on the standard toolbar. A blank publication opens in the Publisher application window.

2. Select the File menu and then Page Setup. The Page Setup dialog box appears.

3. In the dialog box, click the Labels option button in the Choose a Publication Layout box.

4. To select your label type, scroll through the labels listed in the Choose a Label box (see Figure 20.4).

5. When you locate your label number, click the label, and then click OK to exit the dialog box.

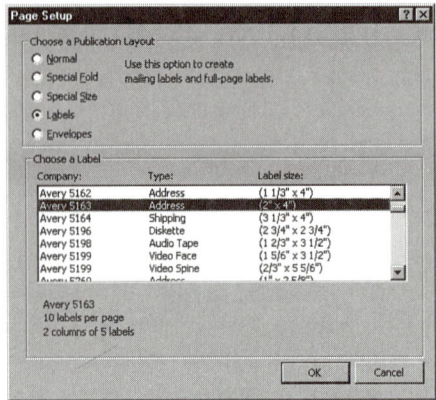

FIGURE 20.4 Select label number in the Page Setup dialog box.

Your new label is available in the Publisher window. You can add text to it as needed by placing a text frame on the label.

In this lesson, you learned about special papers and learned how to create brochures, business cards, envelopes, and mailing labels. In the next lesson, you learn how to create your own Web site using Publisher.

LESSON 21
CREATING A PUBLISHER WEB SITE

In this lesson, you learn to create a Web site for the World Wide Web.

WHAT IS THE WORLD WIDE WEB?

The Internet is a global network of interconnected computer systems that provides several different ways for users to communicate with each other. There is electronic mail (email), Internet Relay Chat, and more importantly, the World Wide Web (WWW). The Web is actually the newest onramp to the Internet and enables you to quickly move from computer to computer (no matter the distance between the computers) by hopping from Web page to Web page using a Web browser (such as Microsoft Internet Explorer).

The Web sites that you view on the Web are made up of linked documents created in the HTML programming language. Your Web browser is able to translate HTML into a format that is viewable (and usable) on your home or office computer.

> **HTML** Hypertext Markup Language is the coding system used to create Web pages for the WWW.

Publisher makes it easy for you to create your own Web site without worrying about HTML coding. You create the Web site in Publisher, and it makes sure that the publication ends up in the correct format. Publisher also enables you to preview your site as you work on it.

CREATING A WEB SITE USING THE WEB SITE WIZARD

Publisher can walk you through the entire Web site creation process. The Web Site Wizard helps you select the various design elements for your new Web site, as well as a page layout. After you create your basic Web site with the wizard, you can modify it using any of Publisher's tools for creating new objects.

To create a Web site using the Web Site Wizard, follow these steps:

1. Click the File menu, and then click New. The Publisher Catalog appears.

2. Click Web Sites in the Wizards pane of the Catalog.

3. Scroll through the layouts available in the Web Sites pane, and select a layout for your Web site (see Figure 21.1).

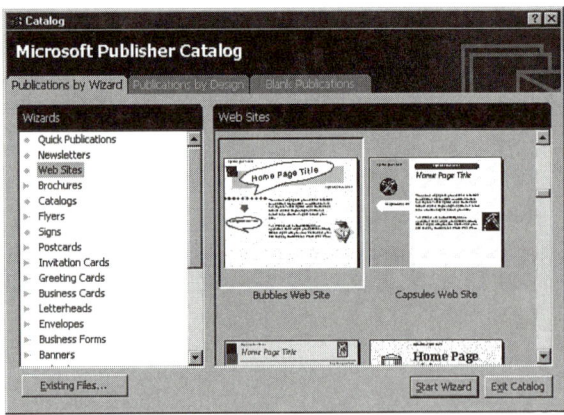

FIGURE 21.1 Select a layout for your Web site in the Publisher Catalog.

4. Click Start Wizard after making your selection. The Web site appears in the publication window, and the wizard appears in the Wizard pane (on the left of the screen).

5. To start the selection of design elements with the help of the wizard, click Next in the Wizard pane.

6. The first choice offered by the Web Site Wizard is the color scheme for the Web site. Select a color scheme (see Figure 21.2), and then click Next to continue.

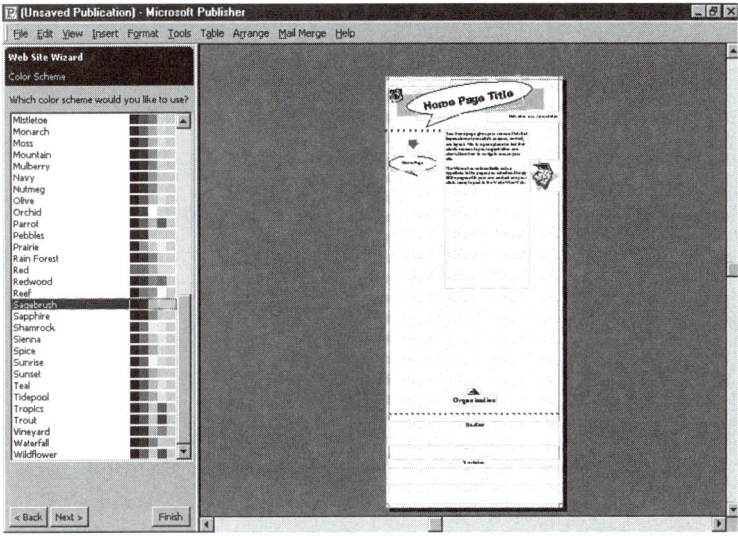

FIGURE 21.2 Select a color scheme for the Web site in the Wizard pane.

7. The next wizard screen provides you with the options of including additional pages in your Web site. You can include a Story page, Calendar page, Event page, and others. You can select some or all the options. Click the checkboxes for the additional pages you want to include.

8. After selecting the additional pages, click Next to continue.

9. The next wizard screen asks if you want to add a form to the Web site. Forms are used to gather information from Web users who visit your site. You can select from the following option buttons: Order form, Response form, Sign-up form, or None (see Figure 21.3). Click your selection.

FIGURE 21.3   You can add user response forms to the Web site.

10. Click Next to continue.

11. The next wizard screen offers you choices related to the navigation bar used on the Web site to link other pages to the site. The navigation bar can be set up with the following choices: both a vertical and horizontal bar, just a vertical bar, or none. Click the appropriate option button.

12. Click Next to continue.

13. The next wizard screen asks you if you want to have a sound played when your Web site opens. Click either Yes or No, and then click Next to continue.

14. The next wizard screen enables you to choose whether a texture (a background pattern) is applied to your Web pages. Select Yes or No, and then click Next to continue.

15. The final wizard screen asks you to select the personal information set to use to replace certain placeholder text on the Web site. Select the information set (see Figure 21.4) by clicking the appropriate option button.

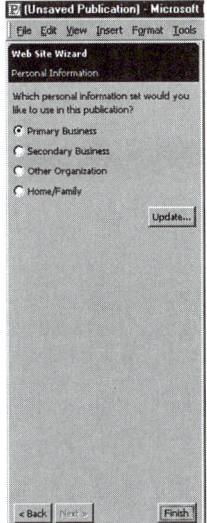

FIGURE 21.4 Choose the personal information set that supplies information for the Web site.

16. Click Finish to complete the Web site.

After you finish working with the wizard, you can edit the Web site pages as you do the pages of any other publication. You can add text (see Lesson 11, "Working with Graphics," for help with text), pictures, and other objects (see Lesson 11 for help adding graphics and Lesson 12, "Adding Special Objects to Your Publications," for help adding special objects). Clip art motion clips are an excellent addition to Web site pages because they provide animated events that add interest to the Web page.

CONVERTING AN EXISTING PUBLICATION TO A WEB SITE

If you prefer to build your Web site without the help of a wizard using the various Publisher features, feel free; you can place text frames, picture frames, and add pages as needed. Publisher can help you convert your pages to the HTML format.

To convert an existing publication to the Web site format, follow these steps:

1. Open the publication you want to convert.

2. Click the File menu, and then click Create Web Site from Current Publication.

3. You are asked if you want to run the Design Checker, as shown in Figure 21.5 (see Lesson 17, "Fine-Tuning Publisher Publications," for information on using the Design Checker). Click No to continue (you can always run the Design Checker after the file has been converted).

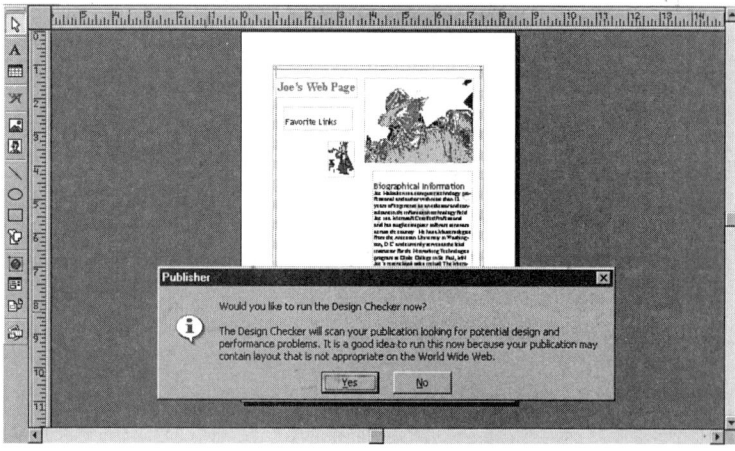

FIGURE 21.5 During the conversion of a publication to a Web site, you are asked to run the Design Checker.

3. After the conversion, you need to save the converted publication. Select the File menu, and then select Save.

4. Type a name for the new Web site, and select a drive and folder for the file. Click Save to complete the process.

After the publication is converted to a Web site, you can preview it in Internet Explorer and save it as a Web site (covered in "Publishing Your Web Site" later in this lesson) for uploading to a Web server.

Adding and Removing Hyperlinks

Web pages are typically linked to other Web pages by hyperlinks.

 Hyperlink A link to another Web site or another page on your own Web site. Hyperlinks can take the form of text entries or pictures. Click them and you are taken to a new location.

Publisher makes it easy for you to insert hyperlinks in your Web site publications. These links become active when you preview your Web site using Microsoft Internet Explorer. Hyperlinks are a great way for you to link multiple pages together for your own Web site or to create links to some of your favorite sites on the Web.

To insert a hyperlink in your publication, follow these steps:

1. Select text in a text frame or another object (such as a picture) on the Web page to serve as the hyperlink (see Figure 21.6).

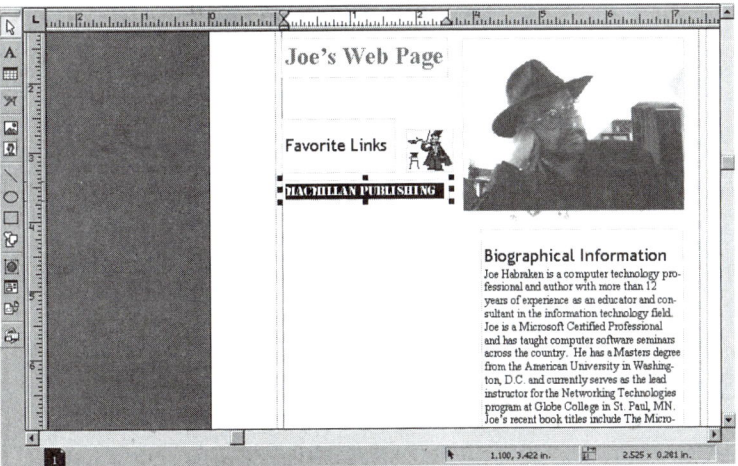

Figure 21.6 Select the text or other object to serve as the hyperlink.

2. Click Insert, and then click Hyperlink. The Hyperlink dialog box appears. You can create hyperlinks to different items using the option buttons at the top of the dialog box.

- **A Web Site or file on the Internet**   Select to create links to other Web sites.

- **Another page in your Web site**   Select to create links to other pages in your own Web site publication.

- **An Internet e-mail address**   Select to create a link that automatically sends an email to the specified address when a user clicks on the link.

- **A file on your hard disk**   Select to create links that enable a user to download a particular file that you specify.

3. To create a link to another Web site, type the address of the Web site that you want to create the link to in the Internet address of the Web site or file box (see Figure 21.7).

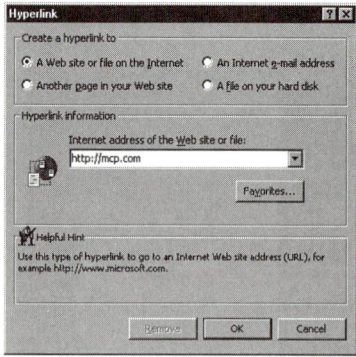

FIGURE 21.7   Specify the Web address for the hyperlink.

4. Click OK. This closes the dialog box and creates the hyperlink in the publication.

CREATING A PUBLISHER WEB SITE   195

Add Hyperlinks from Your Favorites List If you've added Web sites to your favorites list when using the Internet Explorer Web browser, you can add these sites as hyperlinks by clicking the Favorites button in the Hyperlink dialog box. This means you don't have to type the Web site address in the Hyperlink dialog box.

You can also remove a hyperlink from a text entry or a picture on your Web page. This doesn't remove the object, but it does take away the object's linking capabilities.

To remove a hyperlink, follow these steps:

1. Select the text in a text frame or another object on the Web page that contains the hyperlink.

2. Click Insert, and then click Hyperlink. The Hyperlink dialog box appears.

3. Click the Remove button.

VIEWING YOUR WEB SITE

After you set up your Web site publication, you can preview it in the Microsoft Internet Explorer Web browser. This enables you to make sure that your overall design looks good and that items such as hyperlinks work correctly.

To preview your Web site, follow these steps:

1. Click File, and then click Web Page Preview. Internet Explorer opens, showing your Web site (see Figure 21.8).

2. After previewing your Web site, click the Close button to close Internet Explorer.

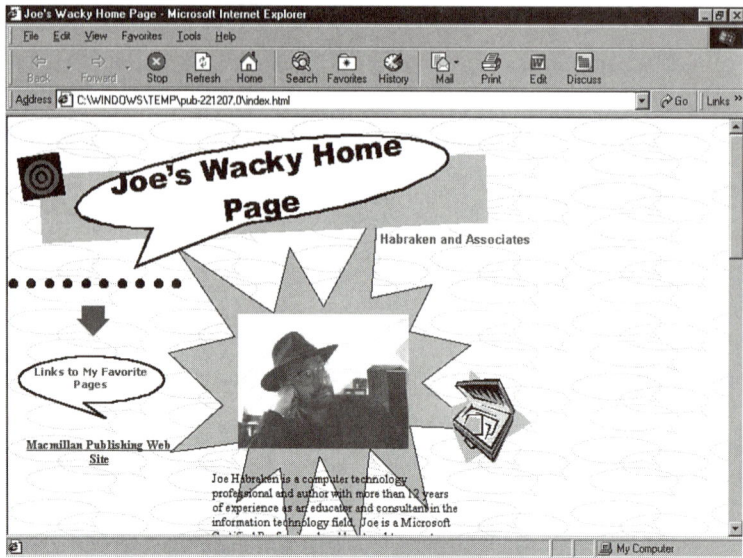

FIGURE 21.8 Preview your Web site in Internet Explorer.

After previewing the site, you return to the Publisher window. You can make changes to the site if needed and preview the site again to see how the changes work.

PUBLISHING YOUR WEB SITE

Web sites that you view on the World Wide Web actually consist of text objects, pictures, and other items that exist as separate elements. The Web page itself is just a way for you (using your Web browser) to link to the different items that you click.

When you create a Web site in Publisher, all the items (the pictures, text, and so on) are saved as part of the publication. For your site to actually work on the Web, the publication has to be broken down into all its elements and saved to a folder. The files in the folder can then be placed on a Web server that enables people on the Web to access your site.

Web Server A computer connected to the Internet that provides access to your Web site.

To publish your Web site, follow these steps:

1. Click File, and then click Save as Web Page. The Save as Web Page dialog box appears.
2. Specify a folder on your computer for the files to be saved in (see Figure 21.9).
3. Click OK. The Web site is saved to the folder as several different component files.

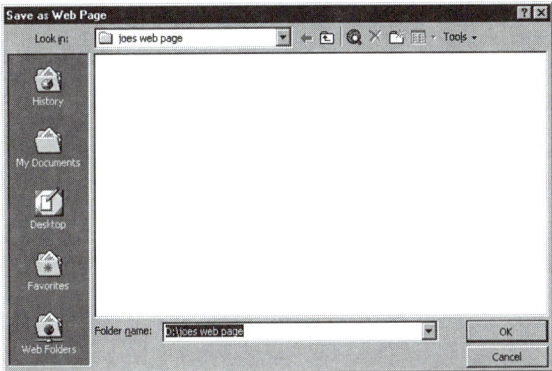

FIGURE 21.9 Specify a folder for the Web site files.

For your Web Site to be available on the World Wide Web, it must be placed on a *Web server*, a computer that "serves up" Web pages to people using a Web browser. Contact your Internet Service Provider for information on uploading your Web page to a Web server.

After you save your Web site to a folder, it is ready to be placed on a Web server and accessed via the World Wide Web. If you decide you want to edit a particular part of the Web site, you need to edit the original Publisher file and then publish the file to a folder as discussed previously. This replaces the original file with the revised file.

In this lesson, you learned to create your own Web site using Publisher. You also learned to insert and remove hyperlinks from your Web site publication. You also learned how to publish your Web site for use on the World Wide Web.

INDEX

A

acquiring graphics, 116-117
adding
    borders, 80-81, 146-148
    bullets, 132-133
    columns and rows, 138-139
    fill colors, 83
    fill effects, 84-85
    mastheads, 94-95
    pages, 57-58
    shadows, frames, 85-86
    text, 87
    see also inserting
Align Objects command, grouping frames, 78
aligning text, 93-94
Alt key, menus, 19
Answer Wizard (help), 9, 65-66
Arrange menu, 78-79
attributes, border, 81-82
audio files, *see* media files
AutoCorrect, 158-160
AutoFormat as You Type tab (AutoCorrect dialog box), 159-160
AutoFormats, formatting tables, 142-143
automatic
    hyphenation, 155-156
    numbering, 131

B

backgrounds, 149
    dates, 150-152
    page numbers, 150-151
    repeating elements, 148
blank publications, 6, 16, 38-41
Border Style dialog box, 80-81
borders, 85
    adding, 80-81
    attributes, 81-82
    colors, 82-83
        changing, 122-123
    pages, adding, 146-148

Bring Forward or Bring to Front (Arrange menu), 78
brochures, trifold, creating, 180-181
bullets
  adding, 132
  automatic, 7
  types of, 133
business cards, creating, 182-183
buttons
  Close, 14
  Control menu, 14
  Gradients, 85
  Maximize/Restore, 14
  Minimize, 14
  Patterns, 85
  Restore/Maximize, 14
  Tints/Shades, 84
  What's This, 68

C

categories of publications, selecting, 27
Cell Diagonals dialog box, 140-141
cells, 134
  dividing, 140-141
  formatting, 139
  merging, 140
  changing
    colors
      borders, 122-123
      fill, 123
      graphics, 106-107
      text, 92-93

fonts, 90
  attributes, 91-92
  Font dialog box, 91-92
  Formatting toolbar, 91
  margins, 145-146
Check Spelling dialog box, 153-155
checking
  AutoCorrect, 158, 160
  design, 156-158
  spelling, 153-155
Clip Gallery, 101-103
  media files, 114-115
Close button, 14
closing publications, 60
Color dialog box, 92-93
Color Matching, 164-165
Color Printing dialog box, 168
color schemes of publications, selecting, 28-29
colors
  borders, 82-83, 122-123
  Color Matching, 164-165
  fill, 83, 123
  graphics, 106-107
  luminescence, 93
  printing, 164-165
  schemes
    professional printing, 167-169
    selecting, 28-29
columns and rows
  adding, 138-139
  sizing, 137-138
commands
  Align Objects, grouping frames, 78
  Save As, 59

Snap to Guide, 49
Zoom, 42, 45-46, 73
completing
   mail merges, 177-178
   publications, design sets, 37
compressing publications, 170-171
connecting, frames, text, 96-98
Contents tab (help), 64-65
context sensitivity, 61
Control menu button, 14
converting publications into Web sites, 191
   Design Checker, 192
copying
   frames, 76
   text, 89
Create Link dialog box, 169-170
Create Table dialog box, 134-135
creating
   business cards, 182-183
   envelopes, 183-184
   frames, *see* frames
   guides, 49-50
   mailing labels, 185-186
   mailing lists, 173, 175
   numbered lists, 131
   publications, 24, 31
      blank publications, 38-41
      design master, 36
      design options, 15-17
      design sets, 34-37
      publication wizards, 25-30, 32-33

trifold brochures, 180-181
Web sites, 187
   converting publications into, 191-192
   Design Checker, 192
   Publisher Catalog, 188-191
   Web Site Wizard, 188-191
cropping graphics, 105-106
custom shapes, drawing, 121

D

data sources, 172
dates, inserting, 150-152
deleting
   frames, 70
   text, 89
Design Checker, 156-158
   converting publications into Web sites, 192
Design Gallery, objects
   editing, 109-110
   inserting, 108-109
design master, 36-37
designs
   blank publications, 38-41
   checking, 156-158
   design master, 36-37
   design sets, 5, 15, 41
      completing a publication, 37
      design master, 36-37
      selecting, 34-36
   options, 17
      blank publications, 16

design sets, 15
Publication Wizard, 15
special paper
  publications, 182
dialog boxes, 21-22
  Auto Correct, 159
  Border Style, 80-81
  Cell Diagonals, 140-141
  Check Spelling, 153-155
  Color, 92-93
  Color Printing, 168
  Create Link, 169-170
  Create Table, 134-135
  Fill effects, 84
  Font, 91-92
  Indents and Lists, 132-133
  Insert Clip Art, 70
  Insert Object, 111-112
  Insert Page, 57-58
  Insert Picture, 100-101
  Layout Guides, 145-146
  Line, 120
  Line Spacing, 127-128
  New Address List, 173
  Nudge, 75
  Open Data Source,
    175-176
  Page Setup
    envelopes, 184
    mailing labels, 185
  Print, 162, 164-165
  Print Settings, Color
    Matching, 164-165
  Print Setup, 163-164
  Recolor Picture, 107
  Save as Web Page, 197
  Size and Position, 73-74

digital cameras, acquiring
  graphics, 116-117
distortion, scaling, 104
dividing cells, 140-141
dragging, sizing objects, 121
drawing objects
  custom shapes, 121
  lines, 119
  Microsoft Draw, 124-126
  ovals, 120-121
  rectangles, 120-121
  tools, 118, 120-122

E

editing
  mailing lists, 175
  special objects
    Design Gallery,
      109-110
    OLE (Object Linking
      and Embedding),
      112-113
  text, 54-55
effects, fill, adding, 84-85
envelopes, creating, 183-184
exiting Publisher, 22-23

F

favorites (Internet Explorer),
  inserting hyperlinks, 195
file formats, graphics, 99-100
files, formats, graphics, 99-100
fill colors
  adding, 83
  changing, 123

fill effects, adding, 84-85
Fill Effects dialog box, 84
flipping objects, 123-124
Font dialog box, 91-92
fonts, 89
    attributes, 91-92
    changing, 90-91
    Color dialog box, 92-93
    Font dialog box, 91-92
formatting
    cells, 139
    numbered lists, 131-132
    objects
        border colors, 122-123
        fill colors, 123
        lines, 120
    paragraphs, 131, 133
        indenting text, 128-129
        line spacing, 127-128
        setting tabs, 129-130
    tables
        AutoFormats, 142-143
        manually, 143-144
Formatting toolbar, 20-21
    alignment, 94
    border attributes, 82
    border color, 82
    fill colors, 83
    font attributes, 91
frames
    borders, 85
        attributes, 81
        Border Style dialog
            box, 80-81
        color, 82-83
    connecting text, 96-98
    copying, 76
    deleting, 70
    fill colors, adding, 83
    fill effects, adding, 84-85
    grouping, 76-78
    inserting, 69-70
    layering, 78-79
    moving
        Nudge dialog box, 75
        Size and Position
            dialog box, 74
    placeholders, 55-57
    shadows, 85-86
    sizing
        handles, 71
        height, 72
        Size and Position
            dialog box, 73
        width, 72
    text, 54

G

Gradients button, 85
graphics
    acquiring, 116-117
    clip art, Clip Gallery,
        101-103
    colors, changing, 106-107
    cropping, 105-106
    Design Gallery, objects
        editing, 109-110
        inserting , 108-109
    file formats, 99-100
    linking (professional
        printing), 169-170
    media files, inserting
        special objects, 113-115

OLE (Object Linking and
Embedding), objects
editing, 112-113
inserting, 111-112
pictures, inserting, 99-101
placeholders, 55-57
scaling, distortion, 104
sizing, 104-105
see also objects
grids, creating, 51
grouping frames, 76-78
guides, 47-48
creating, 49-50
grids, creating, 51

H

height, sizing frames, 72
help, 8
Answer Wizard, 9, 65-66
Contents tab, 64-65
context sensitivity, 61
Index, 66-67
Microsoft Office Web
site, 68
Office Assistant, 8, 15, 61
choosing
characters, 63
searching, 62
window, 63
HTML (Hypertext Markup
Language), 187
see also Web sites
hyperlinks
inserting, 193-195
removing, 195

Hypertext Markup Language
(HTML), 187
see also Web sites
hyphenation
automatic, 155-156
manual, 156

I-J-K

images, *see* graphics
indenting text, 128-129
Indents and Lists dialog box,
132-133
Index (help), 66-67
inkjet printers, 182
Insert Clip Art dialog box, 70
Insert Object dialog box,
111-112
Insert Page dialog box, 57-58
Insert Picture dialog box,
100-101
inserting
clip art, Clip Gallery,
101-103
dates, 150-152
frames, 69-70
hyperlinks, 193-195
merge codes, 175-177
objects, special
Design Gallery,
108-109
media files, 113-115
OLE (Object Linking
and Embedding),
111-112
page numbers, 150-151

pages, 148
pictures, 99-101
tables, 134-135
text, 88
see also adding
Internet, 187
    Web servers, 197
    see also Web sites
Internet Explorer, favorites, inserting hyperlinks, 195

L

launching Publisher, 11, 13
layering
    frames, 78-79
    objects, 122
Layout Guides dialog box, 145-146
layouts
    grids, creating, 51
    guides, 47-50
Line dialog box, 120
lines
    drawing, 119
    formatting, 120
    spacing, 127-128
Lines Spacing dialog box, 127-128
linking graphics (professional printing), 169-170
lists
    bulleted
        adding, 132
        types of, 133

numbered
    creating, 131
    formatting, 131-132
luminescence, 93

M

mail merges, 179
    completing, 177-178
    mailing lists
        creating, 173, 175
        editing, 175
        saving, 174
    merge codes, 172-173
        inserting, 175-177
    previewing, 178
    stopping, 178
mailing labels, creating, 185-186
mailing lists
    creating, 173, 175
    editing, 175
    saving, 174
manual formatting, tables, 143-144
manual hyphenation, 156
margins, changing, 145-146
mastheads, adding, 94-95
Maximize/Restore button, 14
media files
    Clip Gallery, 114-115
    inserting, 113-115
menus, 17, 21
    Alt key, 19
    Arrange, 78-79
    Control menu, 14

menu bar, 14, 18
shortcut menus, 19
View, 42
merges, 179
 cells, 140
 codes, 172-173
  inserting, 175-177
  completing, 177-178
 mailing lists
  creating, 173, 175
  editing, 175
  saving, 174
 previewing, 178
 stopping, 178
Microsoft Draw, 124-126
Office Web site, help, 68
Minimize button, 14
moving
 frames, 74-75
 tables, 136-137

N

New Address List dialog box, 173
new features
 Answer Wizard, 9
 automatic numbering and bullets, 7
 commercial printing support, 9
 Flip Horizontal/Vertical, 9
 help, 8-9
 Office Assistant, 8
 Pack and Go Wizard, 9
 personalized menu/toolbar system, 6
 Save As Web Page, 9
 Web site Wizard, 9
Nudge dialog box, 75
numbered lists
 creating, 131
 formatting, 131-132
numbering
 automatic, 7, 131
 pages, 150-151
numbers, page, inserting, 150-151

O

Object Linking and Embedding, *see* OLE (Object Linking and Embedding)
objects, 20
 drawing
  lines, 119
  Microsoft Draw, 124-126
  ovals, 120-121
  rectangles, 120-121
  tools, 118, 120-122
 flipping, 123-124
 formatting
  border colors, 122-123
  fill colors, 123
  lines, 120
 layering, 122
 rotating, 123
 sizing, dragging, 121
 smart, 108-109

special
   editing, *109-110, 112-113*
   inserting, *108-109, 111-115*
   see also graphics
Office Assistant, 8, 15, 61
   choosing characters, 63
   searching, 62
OLE (Object Linking and Embedding), objects
   editing, 112-113
   inserting, 111-112
Open Data Source dialog box, 175-176
opening publications, 52-53
options, printing, see printing, options
orientation of pages, 29-30, 163-164
ovals, drawing, 120-121

P

Pack and Go (wizard), 170-171
page numbers, inserting, 150-151
Page Setup dialog box
   envelopes, 184
   mailing labels, 185
Page Width view, 43-44
pages
   adding, 57-58
   borders, 146-148
   inserting, 148
   margins, 145-146

paragraphs, formatting, 131, 133
   indenting text, 128-129
   line spacing, 127-128
   setting tabs, 129-130
pasting text, 89
Patterns button, 85
personal profiles, 31-32
personalized menu/toolbar system, 6
pictures, inserting, 99-101
placeholders, 30, 55-57
planning publications, 24-25
previewing
   mail merges, 178
   publications, 161
   Web publications, 161
Print dialog box, 162, 164-165
Print Settings dialog box, Color Matching, 164-165
Print Setup dialog box, 163-164
Print Troubleshooter, 166-167
printing
   inkjet printers, 182
   options
      colors, *164-165*
      orientation of pages, *163-164*
      paper type, *163-164*
      wizards, *164*
   problems, Print Troubleshooter, 166-167
   professionally, 171
      color schemes, *167-169*
      linking graphics, *169-170*
   publications, 162

proportional sizing, 137
Publication
  window, 14
  Wizard, 15
publications
  backgrounds, 149
  dates, 150-152
  page numbers,
    150-151
  repeating
    elements, 148
  categories, selecting, 27
  closing, 60
  color schemes, selecting,
    28-29
  completing, design sets, 37
  compressing, 170-171
  converting into Web sites,
    191-192
  creating, 24
    blank publications,
      38-41
    design master, 36-37
    design options, 15-17
    design sets, 34-37
    publication wizards,
      25-30, 32-33
    selecting design sets,
      34-36
  design sets, 41
  guides, 47-50
  opening, 52-53
  orientation of pages,
    selecting, 29-30
  placeholders, 30
  planning, 24-25
  previewing, 161

print options
  colors, 164-165
  orientation of pages,
    163-164
  paper type, 163-164
printing, 162-163
  professionally, 168-171
rulers, 47-49
saving, 32, 59
special paper
  business cards,
    182-183
  designs, 182
  envelopes, 183-184
  inkjet printers, 182
  mailing lists, 185-186
  Publisher Catalog,
    180-182
  trifold brochures,
    180-181
views
  Page Width, 43-44
  scrolling, 46-47
  Two-Page Spread,
    44-45
  Whole Page, 42-43
Web, previewing, 161
wizards, 33
  categories, 27
  color schemes, 28
  orientation of pages,
    29-30
  personal profiles,
    31-32
  placeholders, 30
  Publisher Catalog,
    25-26

INDEX   209

Publisher Catalog, 25-26
    business cards, 182
    trifold brochures, 180-181
    Web sites, 188-191
Publisher Design Gallery, mastheads, 94-95
Publisher toolbar, 14
    copying, 76
Publisher window, 12
    Close button, 14
    Control menu button, 14
    menu bar, 14
    Minimize button, 14
    Office Assistant, 15
    Publication window, 14
    Publisher toolbar, 14
    Quick Publication Wizard, 15
    Restore/Maximize button, 14
    Scrollbar, 15
    Standard toolbar, 14
    Status bar, 14
    Title bar, 13
publishing Web sites, 196-197

Q-R

Quick Publication Wizard, 15

Recolor Picture dialog box, 107
rectangles, drawing, 120-121
repeating elements, 148
Restore/Maximize button, 14
rotating objects, 123
rows and columns
    adding, 138-139
    sizing, 137-138

rulers, 47-49
    setting tabs, 130

S

Save as Web Page dialog box, 197
saving
    mailing lists, 174
    publications, 32, 59
scaling, *see* sizing, scaling
scanners, acquiring graphics, 116-117
Scrollbar, 15
scrolling, 46-47
searching
    for clip art, 103
    Office Assistant, 62
selecting
    design sets, 34-36
    orientation of pages, 29-30
    publications
        categories, 27
        color schemes, 28-29
    text, 88-89
Send Backward (Arrange menu), 79
Send to Back (Arrange menu), 78
servers, Web, 197
setting tabs
    ruler, 130
    Tab selector, 129
shadows, frame, 85-86
shortcut menus, 19
sites, Web, *see* Web sites

Size and Position dialog box, 73-74
sizing
    columns and rows, 137-138
    frames, 71
        height, 72
        Size and Position dialog box, 73
        width, 72
    graphics, 104-105
    handles, frames, 71
    objects, dragging, 121
    proportional, 137
    scaling, distortion, 104
    tables, 135-136
smart objects, 108-109
Snap to Grid feature, 75
special objects
    editing
        Design Gallery, 109-110
        OLE (Object Linking and Embedding), 112-113
    inserting
        Design Gallery, 108-109
        media files, 113-115
        OLE (Object Linking and Embedding), 111-112
special paper publications
    business cards, 182-183
    design, 182
    envelopes, 183-184
    inkjet printers, 182
    mailing labels, 185-186
    Publisher Catalog, 180-182
    trifold brochures, 180-181
Spell Checker, 153-155
Standard toolbar, 14, 20-21
starting Publisher, 11, 13
Status bar, 14

T-U

Tab selector, 129
tables
    cells, 134
        dividing, 140-141
        formatting, 139
        merging, 140
    columns and rows
        adding, 138-139
        sizing, 137-138
    cutting and pasting, 137
    formatting
        AutoFormats, 142-143
        manually, 143-144
    inserting, 134-135
    moving, 136-137
    sizing, 135-136
    text, entering, 141-142
tabs, setting
    ruler, 130
    Tab selector, 129
text
    adding, 87
    alignment, 93-94
    colors, 92-93
    copying, 89

deleting, 89
editing, 54-55
entering in tables, 141-142
fonts, 89
  attributes, 91-92
  changing, 90-91
hyphenation
  automatic, 155-156
  manual, 156
indenting, 128-129
inserting, 88
mastheads, adding, 94-95
pasting, 89
paragraphs, formatting, 131, 133
  indenting text, 128-129
  line spacing, 127-128
  setting tabs, 129-130
selecting, 88-89
spell checking, 153-155
text frames, 54
  connecting, 96-98
Tints/Shades button, 84
Title bar, 13
toolbars, 17
  formatting, 20-21
  Formatting,
    alignment, 94
    border attributes, 82
    border color, 82
    fill colors, 83
    font attributes, 91
  Publisher, 14
    copying, 76
  Standard, 14, 20-21
  Status, 14

trifold brochures, creating, 180-181
troubleshooting, Print Troubleshooter, 166-167
Two-Page Spread, 44-45
typefaces, *see* fonts

V

video files, *see* media files
viewing
  grids, 51
  guides, 47-50
  Page Width, 43-44
  rulers, 47-49
  scrolling, 46-47
  Two-Page Spread, 44-45
  Web sites, 195-196
  Whole Page, 42-43

W-X-Y-Z

Web pages, *see* Web sites
Web publications, previewing, 161
Web servers, 197
Web sites
  converting publications into, 191-192
  creating, 187-191
  hyperlinks
    inserting, 193-195
    removing, 195
  Microsoft Office, 68
  publishing, 196-197

viewing, 195-196
Web servers, 197
Web site Wizard, 188-191
What's This button, 68
Whole Page view, 42-43
widths, sizing frames, 72
window, Publisher, 12
   Close button, 14
   Control menu button, 14
   menu bar, 14
   Minimize button, 14
   Office Assistant, 15
   Publication window, 14
   Publisher toolbar, 14
   Quick Publication
    Wizard, 15
   Restore/Maximize
    button, 14
   Scrollbar, 15
   Standard toolbar, 14
   Status bar, 14
   Title bar, 13
windows, help, 63
wizards, 5
   Answer, 9
   margins, 145
   Pack and Go, 170-171
   page borders, adding,
    146-148
   print options, 164
   Web Site, 188-191
WWW (World Wide Web), *see*
   Web sites

Zoom, 42, 46
Zoom command, 45-46, 73

Turn to the *Authoritative* Encyclopedia of Computing

You'll find over 150 full text books online, hundreds of shareware/freeware applications, online computing classes and 10 computing resource centers full of expert advice from the editors and publishers of:

- Adobe Press
- BradyGAMES
- Cisco Press
- Hayden Books
- Lycos Press
- New Riders
- Que
- Que Education & Training
- Sams Publishing
- Waite Group Press
- Ziff-Davis Press

mcp.com
The Authoritative Encyclopedia of Computing

When you're looking for computing information, consult the authority. The Authoritative Encyclopedia of Computing at mcp.com.

Get the best information and learn about latest developments in:

- Design
- Graphics and Multimedia
- Enterprise Computing and DBMS
- General Internet Information
- Operating Systems
- Networking and Hardware
- PC and Video Gaming
- Productivity Applications
- Programming
- Web Programming and Administration
- Web Publishing

SAMS Teach Yourself in 10 Minutes

Quick steps for fast results™

Sams Teach Yourself in 10 Minutes *gets you the results you want—fast! Work through the 10-minute lessons and learn everything you need to know quickly and easily. It's the handiest resource for the information you're looking for.*

Word 2000
Peter Aitken
ISBN: 0-672-31441-x
$12.99 US/$18.95 CAN

Other Sams Teach Yourself in 10 Minutes Titles

Powerpoint 2000
Faithe Wempen
ISBN: 0-672-31440-1
$12.99 US/$18.95 CAN

Excel 2000
Jennifer Fulton
ISBN: 0-672-31457-6
$12.99 US/$18.95 CAN

Outlook 2000
Joe Habraken
ISBN: 0-672-31450-9
$12.99 US/$18.95 CAN

Access 2000
Faithe Wempen
ISBN: 0-672-31487-8
$12.99 US/$18.95 CAN

Office 2000
Laura Acklen
ISBN: 0-672-31431-2
$12.99 US/$18.95 CAN

Windows 98
Jennifer Fulton
ISBN: 0-672-31330-8
$12.99 US/$18.95 CAN

Internet Explorer 5
Jill Freeze
ISBN: 0-672-31646-3
$12.99 US/$18.95 CAN

Frontpage 2000
Galen Grimes
ISBN: 0-672-31498-3
$12.99 US/$18.95 CAN

Visual Basic 6
Lowell Mauer, et.al
ISBN: 0-672-31458-4
$12.99 US/$18.95 CAN

All prices are subject to change.

SAMS

www.samspublishing.com

Get FREE books and more...when you register this book online for our Personal Bookshelf Program

http://register.samspublishing.com/

SAMS

Register online and you can sign up for our *FREE Personal Bookshelf Program*...unlimited access to the electronic version of more than 200 complete computer books—immediately! That means you'll have 100,000 pages of valuable information onscreen, at your fingertips!

Plus, you can access product support, including complimentary downloads, technical support files, book-focused links, companion Web sites, author sites, and more!

And, don't miss out on the opportunity to sign up for a *FREE subscription to a weekly email newsletter* to help you stay current with news, announcements, sample book chapters, and special events, including sweepstakes, contests, and various product giveaways.

We value your comments! Best of all, the entire registration process takes only a few minutes to complete...so go online and get the greatest value going—absolutely FREE!

Don't Miss Out On This Great Opportunity!

Sams is a brand of Macmillan Computer Publishing USA
For more information, please visit *www.mcp.com*

Copyright ©1999 Macmillan Computer Publishing USA